6th International Conference on Material Science and Engineering Technology (ICMSET 2022)

IOP Conference Series: Materials Science and Engineering
Volume 1250

Online
25 - 26 March 2022

ISBN: 978-1-7138-6016-7
ISSN: 1757-8981

Printed from e-media with permission by:

Curran Associates, Inc.
57 Morehouse Lane
Red Hook, NY 12571

Some format issues inherent in the e-media version may also appear in this print version.

Institute of Physics
Dirac House, Temple Back
Bristol BS1 6BE UK

Phone: 44 1 17 929 7481
Fax: 44 1 17 920 0979

techtracking@iop.org

Additional copies of this publication are available from:

Curran Associates, Inc.
57 Morehouse Lane
Red Hook, NY 12571 USA
Phone: 845-758-0400
Fax: 845-758-2633
Email: curran@proceedings.com
Web: www.proceedings.com

TABLE OF CONTENTS

Preface

The ICMSET 2022 is the 6th International Conference on Material Science and Engineering Technology (6th ICMSET 2022) was planned in Jeju Island, Korea during March 25-27, 2022. But currently, the entire world is struggling against the virulent pandemic COVID-19, so that the ICMSET 2022 conference changed the presentation method from Oral presentation to Remote Video Presentation to help with Global Quarantine Activities.

The Virtual of ICMSET 2022 was held on Incheon National University at Video Conference Meeting Room on March 25-27, 2022. The ICMSET 2022 Committee members attended conference at on-site and also authors attended the conference as Video Meeting. The conference program covered Invited Keynote Speech, Oral, and Poster Presentations, they presented their presentation 15 minutes including Q&A time.

The committee of ICMSET 2022 expresses their sincere thanks to all authors for their high-quality research papers and careful presentations. All reviewers are also thanked for their careful comments and advices. Thanks, are finally given to IOP Publication as well for producing this volume.

The Organizing Committee of ICMSET 2022

Committee Chair

Prof. Mosbeh Kaloop

Incheon Disaster Prevention Research Center

Incheon National University

IOP Conf. Series: Materials Science and Engineering 1250 (2022) 011001 doi:10.1088/1757-899X/1250/1/011001

Sponsors

Incheon Disaster Prevention Research Center (IDPRC)

Recently, various efforts to prevent and prepare are vitally needed for prevention of Disasters and calamities. So we understand the necessity for technology of disaster and we built up the Incheon Disaster Prevention Research Center (IDPRC) in Incheon National University (INU) in 1997.

Accordingly, Incheon Disaster Prevention Research Center (IDPRC) in Incheon National University has progressed of research on the prevention of disasters and calamities through the various seminars, conference and lectures. These research could be conducted cooperation with Incheon National University (INU) in various fields' structure, soil, hydraulic and environment.

Incheon Disaster Prevention Research Center (IDPRC) will try to be a leader in the disaster of industry through the various research activities and global conference.

IOP Conf. Series: Materials Science and Engineering 1250 (2022) 011001 doi:10.1088/1757-899X/1250/1/011001

Lists of Committees

Committee name	Organizer name	Organization
Honorary Chair	Mosbeh Kaloop	Incheon National University
Conference General Chairs	Mosbeh Kaloop	Incheon National University
Honorary Chair	Dongkeon Kim	Dong-A University
Technical Committees	J. W. Hu	Incheon National University

IOP Conf. Series: Materials Science and Engineering 1250 (2022) 011002 doi:10.1088/1757-899X/1250/1/011002

Peer Review Statement

All papers published in this volume have been reviewed through processes administered by the Editors. Reviews were conducted by expert referees to the professional and scientific standards expected of a proceedings journal published by IOP Publishing Publishing.

- **Type of peer review:** Double Anonymous
- **Conference submission management system:** Morressier
- **Number of submissions received:** 50
- **Number of submissions sent for review:** 38
- **Number of submissions accepted:** 10
- **Acceptance Rate (Submissions Accepted / Submissions Received × 100):** 20
- **Average number of reviews per paper:** 3
- **Total number of reviewers involved:** 5
- **Contact person for queries:**
  Name: R. Kaloop Mosbeh
  Email: mosbeh@inu.ac.kr
  Affiliation: Incheon Disaster Prevention Research Center

IOP Conf. Series: Materials Science and Engineering 1250 (2022) 012001 doi:10.1088/1757-899X/1250/1/012001

Orbital Change of belt for Metal V-belt Type CVT (Continuously Variable Transmission) - Change of pitch angle at strings -

Kazuki Oshima [1], Seiji Mitsubayashi [1], Kiyotaka Obunai [2] Kazuya Okubo [2]

[1] Graduate School of Doshisha University, Engineering 1-3 Tataramiyakodani, Kyotanabe, Kyoto, 610-0394, Japan

[2] Doshisha University, Engineering 1-3 Tataramiyakodani, Kyotanabe, Kyoto, 610-0394, Japan

ctwf0561@mail4.doshisha.ac.jp

Abstract. The object of this study is to investigate the orbital change of belt with the increase of the transfer torque, during steady-state belt of rotation using a bench-type CVT testing machine. Acceleration sensors were attached to the V-belt to measure accelerations of each element. The pitch angle of each element at the strings also near the entrances and exit was continuously measured by calculating the 3-axis accelerations during the belt travel. Macroscopic change of pitch angle was almost followed by the geometrical change of the angle during the orbital motion of the belt. However, local change of pitch angle at near the entrance and exit of the pulley was not explained by the angle of the geometric posture movement of elements. The pitch angle of each element was locally increased near the entrance and exit of the driven pulley as the load torque increased, while the opposite tendency was observed near those locations of the driving pulley.

1. Introduction

Today, there is a growing demand in the automotive industry for fuel efficiency reduction from the viewpoint of environmental protection and the depletion of fossil fuels. In order to meet these demands, automakers are working to improve the efficiency of powertrains such as transmissions [1]. Among them, the V-belt type CVT (Continuously Variable Transmission), a type of transmission, has become the mainstream of transmissions in Japan. There are five major types of transmissions for automobiles. Automatic transmissions are classified into two types: gear transmissions, which transmit power by means of gears and change gears in a staircase manner, and friction transmissions, which transmit power by means of the frictional force acting between the belt and pulley and change gears continuously. As shown in Figure 1, CVT is a transmission which takes an advantage of frictional mechanism and consisting of a driving V-pulley, driven V-pulley and a V-belt. The V-groove width can be changed by applying hydraulic pressure to the cylinder, and the wrapping diameter can be continuously varied. Unlike the step automatic transmission (AT), the vehicle speed can be changed sequentially without significantly changing the engine speed when the vehicle accelerates or slow down, while maintaining the most thermally efficient operating point of the engine, thus enabling operation with good fuel efficiency and power performance [2]. However, frictional transmission requires a much larger pressing load than that for gear transmission, resulting in higher transmission losses. Under such severe conditions, it is necessary to clarify the behavior of elements that support large pressing loads in order to meet the demand for low fuel consumption in automobiles [3]. However, the belt deflection of strings during belt traveling has not been clarified, while behavior at near the front and rear strings in the pulley groove is also important in order to clarify elements. The object of this study is to investigate the orbital change of belt with the increase of the transfer torque, during steady-state belt of rotation using a bench-type CVT testing machine.

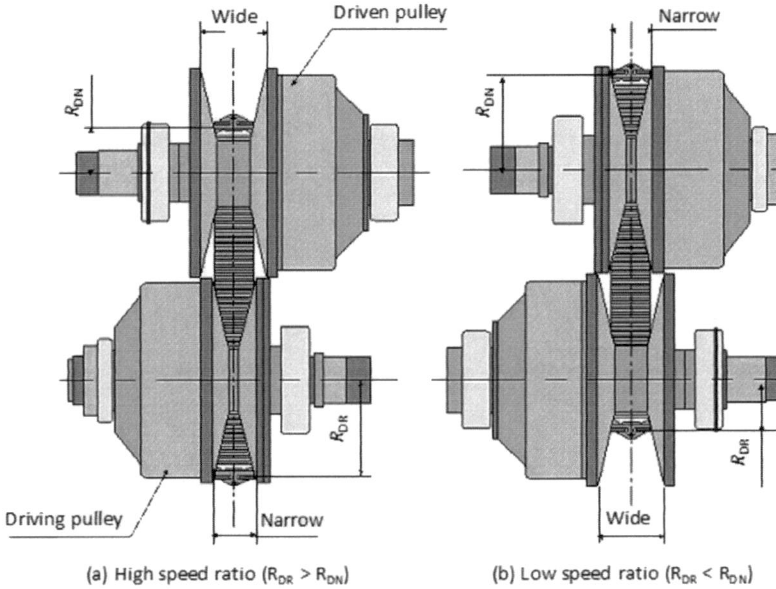

Figure 1. CVT set-up.

2. Test method

2.1. Testing machine

The appearance of the testing machine used in this study is shown in Figure 2, and its schematic diagram is shown in Figure 3. The testing machine consists of a driving and driven pulley with variable groove widths, an A/C motor, two torque meters, a load cell for measuring the inter-shaft force, and an electromagnetic brake. In this system, the power generated by the A/C motor was transmitted from the driving shaft to the driven shaft through the CVT unit, and the power from the A/C motor was absorbed by the electromagnetic brake. The clamping forces on driving and driven pulley were applied by hydraulic pressure. Lubrication with CVT fluid was applied to each contact area in the CVT unit to reduce wear due to contact between metal surfaces between the belt and pulley. Cylindrical steel collar was inserted between back face of the driving pulley and bearing wall to keep the speed ratio constant.

Figure 2. Appearance of testing machine.

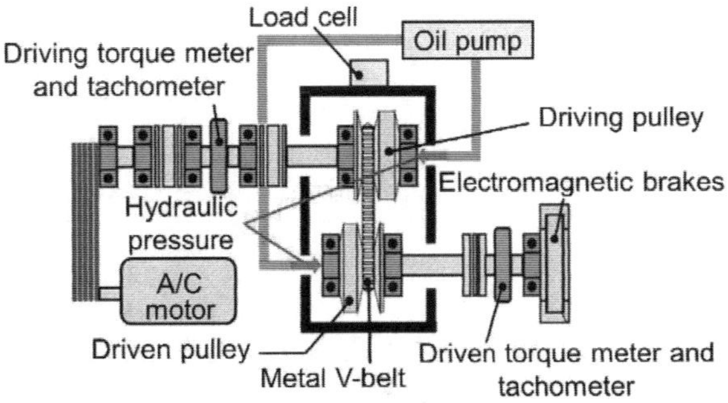

Figure 3. Schematic view of testing machine.

2.2. *Elements for measuring accelerations*

Specific element was prepared to measure the belt accelerations. Acceleration data were acquired using two types of sensors: a low-noise 3-axis accelerations sensor (Analog Devices ADXL355) and an optical sensor (CdS cell). The sensors used for the measurements are shown in Figure 4, and the directions of accelerations were defined as shown in Figure 5. The sensors were placed in a case made with plastic resin by a 3D printer. The case was fixed to the top of a T-shaped steel plate which was brazed to the elements of the belt to join it rigidly. This made it possible to carry out measurements while maintaining the transmission state during drive.

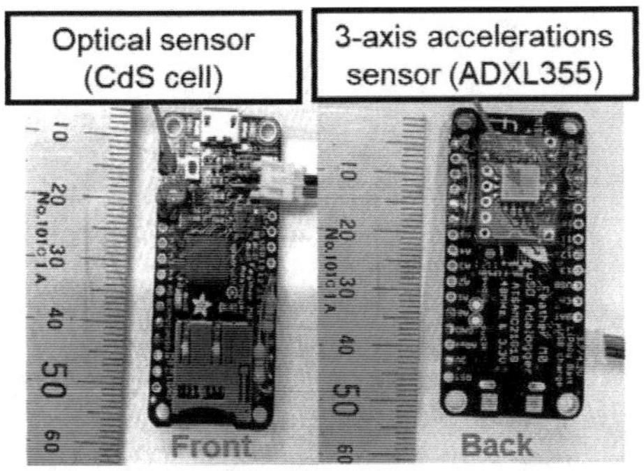

Figure 4. Magnified view of sensors.

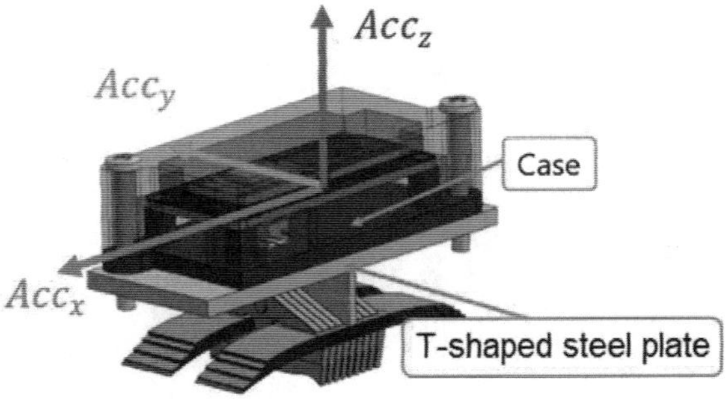

Figure 5. Definition of measuring directions.

2.3. *Measurements of element accelerations and calculation of pitch angle*

Specific element introduced in section 3.2 were attached to the belt is shown in Fig. 6, and acceleration data obtained from the sensors were recorded on a microSD card during running of the belt. The experimental conditions are shown in Table 1. The pitch angle $p[r\ a\ d]$ can be calculated by following equation with the 3-axis accelerations data $A\ c\ c_{x,y,z}[m/s^2]$ obtained under each condition.

$$p = \sin^{-1}\left(\frac{-A\ c\ c_x}{\sqrt{9.81^2 - Acc_y{}^2}}\right) \quad (1)$$

Measurements were started when the sensors were located at the position of entrance of the driven pulley.

IOP Conf. Series: Materials Science and Engineering 1250 (2022) 012001 doi:10.1088/1757-899X/1250/1/012001

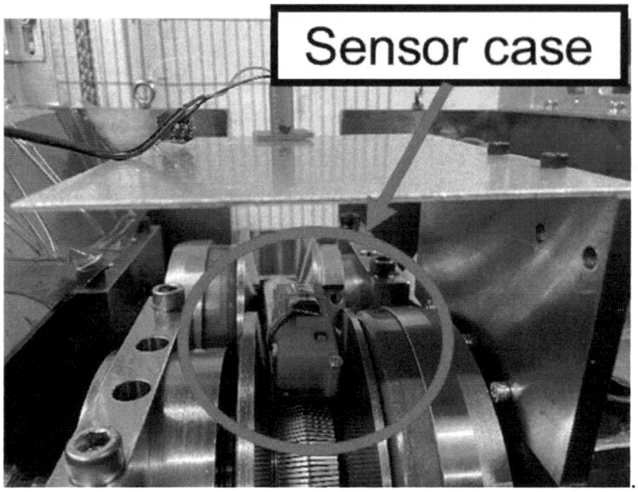

Figure 6. Acceleration sensor attached to belt.

Table 1. Test conditions.

Driving rotational speed [rpm]	5
Speed ratio [-]	1
Driving clamping force [kN]	7.4
Driven clamping force [kN]	6.3
Torque [Nm]	0, 10, 20, 30, 40, 50

3. Test results

The change of acceleration of the element in the direction of belt travel is shown in Figure 7. The change in pitch angle calculated by equation (1) is shown in Figure 8, where the pitch angle was described in global co-ordinate. The test results showed that macroscopic change of pitch angle was almost followed by the geometrical change of the angle during the orbital motion of the belt. However, local change of pitch angle at near the entrance and exit of the pulley was not explained by the angle of the geometric posture movement of elements. This change was depended on the condition of applied torque. The pitch angle of each element was locally increased near the entrance and exit of the driven pulley as the load torque increased, while the opposite tendency was observed near those locations of the driving pulley. These results suggested that the elements were slightly moved to outward from theoretical line of the circumference at the intensive compression string, while they were moved to inward slightly at extensive compression string of the belt.

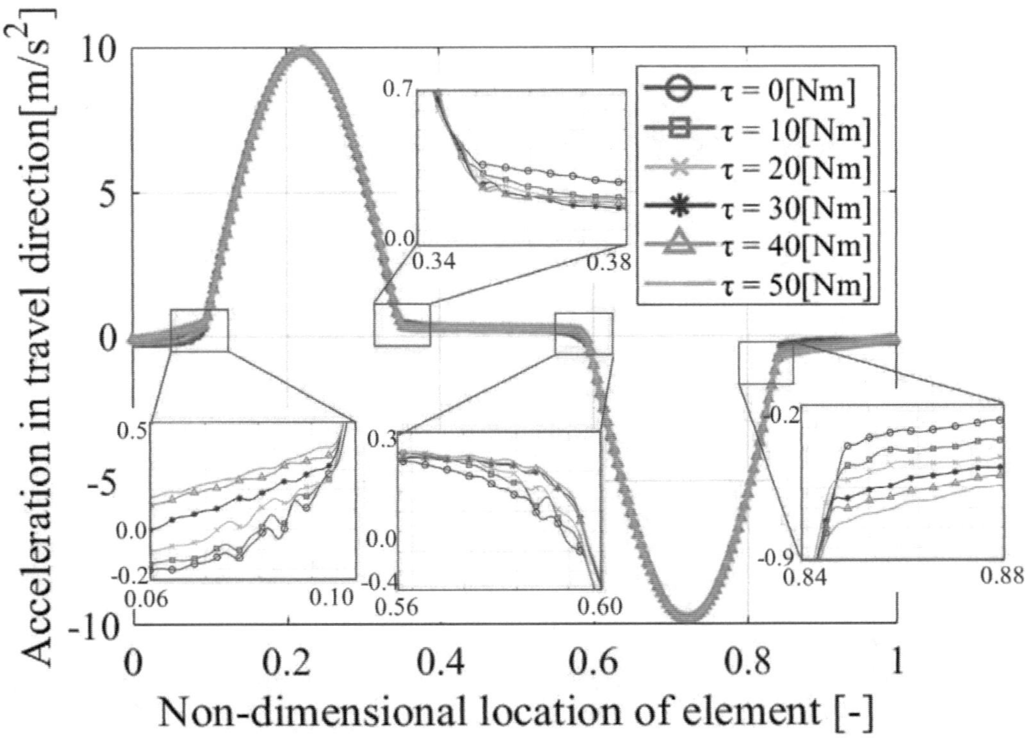

Figure 7. Acceleration of element in tangential direction.

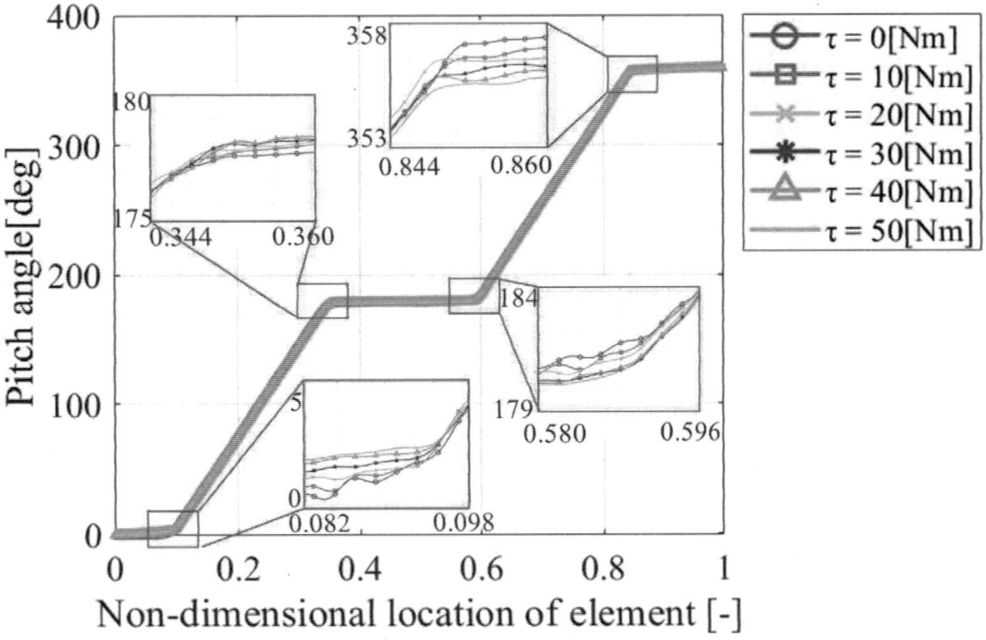

Figure 8. Change of pitch angle.

4. Conclusion

- Macroscopic change of pitch angle was almost followed by the geometrical change of the angle during the orbital motion of the belt.
- Local change of pitch angle at near the entrance and exit of the pulley was not explained by the angle of the geometric posture movement of elements.
- The pitch angle of each element was locally increased near the entrance and exit of the driven pulley as the load torque increased, while the opposite tendency was observed near those locations of the driving pulley.
- The elements were slightly moved to outward from theoretical line of the circumference at the intensive compression string, while they were moved to inward slightly at extensive compression string of the belt.

References

[1] Ahmet Yildiz, Francesco Bottiglione, Antonio Piccininni, Osman Kopmaz, Giuseppe Carbone, "Experimental validation of the Carbone – Mangialardi - Mantriota model of continuously variable transmissions", Journal of Automobile Engineering, Institution of Mechanical Engineers 2018, vol.232, p.828-837

[2] Lingyuan Kong, Robert G. Parker, "Steady mechanics of layered, multi-band belt drives used in continuously variable transmissions (CVT)" Mechanism and Machine Theory 43 (2008), 171-185

[3] Nilabh Srivastava, Imtiaz Haque, "A review on belt and chain continuously variable transmissions (CVT): Dynamics and control", Mechanism and Machine Theory 44 (2009), 19-41

Acknowledgments

This research was commissioned by the Research Association for Automotive Power Transmission Technology (TRAMI) under the "FY2021 Project for Establishing a Simulation Infrastructure for Accelerating the Development of Next-Generation Vehicles.

IOP Conf. Series: Materials Science and Engineering 1250 (2022) 012002 doi:10.1088/1757-899X/1250/1/012002

Environmental model using life cycle assessment and artificial intelligence techniques to predict impacts on industrial water treatment.

José de Jesus[1], Karla Oliveira-Esquerre[1], Diego Lima Medeiros[1]

[1]Federal University of Bahia (UFBA), Growing with Applied Metrics and Mindful Analysis (GAMMA), Industrial Engineering Program (PEI). Professor Aristídes Novis Street, 02, Federação, 40210-630, Salvador, Bahia, Brazil. *Corresponding author:

jose.oduque@hotmail.com

Abstract. With the advancement of globalization, the growing demand for environmental resources in industrial processes and the increasing availability of data and information, the need to align data modeling concepts with Artificial Intelligence (AI) techniques and existing environmental tools has emerged. From a sustainability perspective, life cycle assessment (LCA) is an extremely important tool in ensuring adequate practices in environmental thinking. It is through the life cycle assessment (LCA) that it is possible to measure the environmental impacts from products and processes, as well as to make projections that minimize these impacts. This research employed an artificial intelligence (AI) methods, namely adaptive neurofuzzy inference system (ANFIS) model, to predict life cycle environmental impacts in industrial water treatment using aluminum sulfate and Tannin-Base biocoagulant. The results show that different AI algorithms are used to build LCA models. The AI algorithms in the studies work from problem identification to the solution stage, so the integration between AI and LCA makes it possible to build predictive machine learning models to enable assertiveness in decision making.

1. Introduction

Currently, the growing environmental problem has been causing the awakening of society's sustainable awareness and providing a greater movement in the effort to ensure harmony between the economy, growth and environmental conservation. In this context, sustainable development plays an extremely important role in the balance between society's productivity demands and the preservation of existing natural resource reserves. From a sustainability perspective, life cycle assessment (LCA) is an extremely important tool in ensuring adequate practices in environmental thinking. It is through life cycle assessment (LCA) that it is possible to measure the environmental impacts from products and processes, as well as to make projections that minimize these impacts, however with the increasing availability of data and information on products and processes, making these systems increasingly increasingly complex and automated, you have to consider that there are challenges with regard to the four steps of the LCA (i.e. objective and scope definition, lifecycle inventory, lifecycle impact analysis and interpretation) that must keep up with the trend of application of the tool and find a combination of methods that allows the evaluation and provides a perspective of promising research directions ensuring assertiveness in decision making. The volume of data and information generated daily in commercial transactions, social networks, analyses, sales and delivery of products, logistical processes, search engines and others, makes the use of Artificial Intelligence systems part of society's daily life. Modern. These systems are able to extract accurate and valuable information, perform tasks, solve problems and make sense of data from their analysis. The use of artificial intelligence systems has contributed to the solution of problems in various areas of society, especially with regard to environmental issues, such as real-time monitoring of devastated areas, studies of the dimension of environmental impacts, mapping of regions affected by ecological disasters, minimization of uncertainties associated with environmental studies, and others. The main advantage of using artificial intelligence systems is that the community needs to have a thorough understanding of the relevant technologies behind modern data science, and it also needs to form a strong vision of what can be done with these new technologies in the future. context of environmental applications. The concepts of Artificial Intelligence (AI), Big Data and Machine Learning, for example, have been worked on by data scientists around the world and transformed many aspects of political, social and economic dimensions. Data scientists argue that emerging data science should be taken as a new paradigm of science and research, and that it still requires a long evolution from empirical observation, theoretical analysis and computer simulations. With this, it is necessary to see data science as a new field for research, taking advantage of all the availability of data with more advanced techniques, of artificial intelligence, to synthesize and examine existing theories and models. Data analysis has always been one of the main mechanisms of scientific investigation, but nowadays, scientists need to lean between the new generation and data analysis, with the classicist methods of existing analysis. Therefore, the main objective of this research is to conduct an evidence-based review, using data mining techniques on the context of Life Cycle Assessment (LCA) and its intersection with artificial intelligence methods in different sectors of society, as well as, understand its main technological and institutional obstacles to the adoption of these technologies in the world scenario.

2. Materials and methods

In this study, the ANFIS method was used to model the production of tannin-based biocoglante, described below in Figure 1 and Table 1, Table 2 and Table 3.

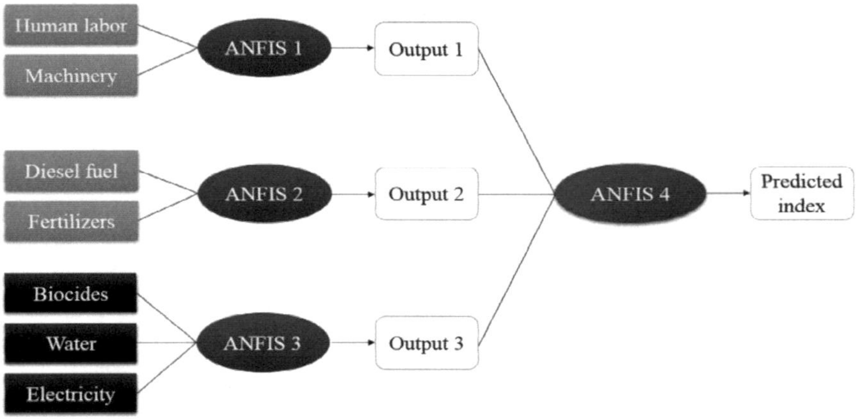

Figure. 1. Two-level ANFIS structure for environmental prediction of water treatment using aluminum sulfate and tannin-based biocoagulant.

Fuzzy logic, as one of AI methods, provides a numeric system and al- lows for simple presentation of production process knowledge by IF- THEN rules (Singh and Gill, 2010). Fuzzy set theory and the associated fuzzy logic are considered a powerful and well-known modeling technique (Zedeh, 1965).

Based on Sugeno fuzzy modeling, natural language is employed for qualitative action of the system (Sugeno and Yasukawa, 1993).

Quantities of energy inputs and outputs in the production of biocoagulants and use in water treatment.

Table 1 Gate-to-gate inventory of the production of 1 kg of tannin-based biocoagulant.

Materials	Amount	Unit	*GSD2
Acacia Bark	29,44	kg	1,07
Water	0,87	m^3	1,05
Electricity	0,63	kWh	1,5
Methanol	3,25	L	1,05

Source: Own elaboration
*GSD2: square geometric standard deviation.

Table 2 Foreground inventory for 1 m^3 of distributed water.

Stage	Amount	Unit	*GSD2
Uptake			
Chlorine	0.030	kg	1.05
Electricity	0.54	kWh	1.07
Treatment			
Aluminum Sulfate	0.26	kg	1.05
Biocoagulant	0,08	kg	1.05
Sodium Carbonate	0.12	kg	1.05
Polyelectrolyte	0.0026	kg	1.05
Electricity	0.38	kWh	1.07
Chlorine	0.025	kg	1.05
Sludge	0.020	m3	1.09
Infrastructure	3.07E-9	unit	3.28
Sludge generated			
	0,02	m^3	1,05

*GSD2: square geometric standard deviation.

Table 3 Emissions from biocoagulant production.

Materials	Amount	Unit	*GSD2
Emissions to Air			
Methanol	2,57	kg	1,50
Water	0,23	kg	1,05
Particulate matter	0,07	kg	3,00
Total organic carbon	0,007	kg	1,05
Organic compounds volatile	0,003		1,50
Carbon dioxide, biogenic	0,36	kg	1,05
Sulfur dioxide	0,04	kg	1,05
Carbon monoxide	0,18	kg	1,05
Emissions for Treatment			
Residual waters	0,8	m^3	1,50

*GSD2: square geometric standard deviation.

3. Results and Discussion

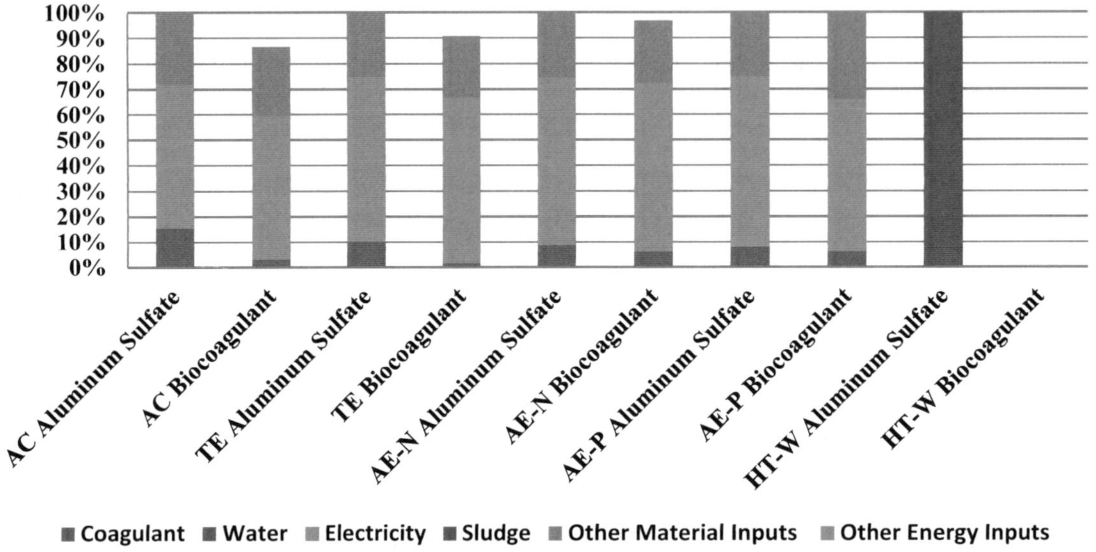

Figure 2. shows the results of this contribution.

The environmental impact contribution process aims to unify the results obtained from different categories of environmental impacts on the same graph and magnitude. The coagulant that presented the greatest contributions within the human categories was aluminum sulfate, with emphasis on the categories of ecotoxicity and toxicity. In the other categories, observe the environmental performance of aluminum superiors to the environmental impacts in relation to the biocoagulant. In this way, the environment remains the toxicity of the human ecotoxicity categories with regard to the life cycle of the use of coagulating agents in the outstanding processes of industrial water treatment. In order to identify an energy demand, water footprint and carbon footprint of the evaluated process, contributions to these categories were also simulated.

Table 4 ANFIS structure to eliminate the environmental impacts of water treatment using aluminum sulfate and tannin-based biocoagulant.

Category	ANFIS models	Type		Number		Learning method	R^2	RMSE	MAPE (%)
		Input	Output	Input	Epoch				
	ANFIS (1)	Gbell	Linear	5,6	32	Hybrid	0.409	0.652	0.080
	ANFIS (2)	Gbell	Linear	5,6	32	Hybrid	0.472	0.633	0.091
	ANFIS (3)	Gbell	Linear	5,6	32	Hybrid	0.342	0.679	0.091
AC	ANFIS (4)	Gbell	Linear	5,6	32	Hybrid	0.454	0.528	0.057
	ANFIS (1)	Gbell	Linear	5,6	32	Hybrid	0.293	0.571	0.018
	ANFIS (2)	Gbell	Linear	5,6	32	Hybrid	0.384	0.641	0.087
	ANFIS (3)	Gbell	Linear	5,6	32	Hybrid	0.329	0.559	0.082
TE	ANFIS (4)	Gbell	Linear	5,6	32	Hybrid	0.331	0.584	0.056
	ANFIS (1)	Gbell	Linear	5,6	32	Hybrid	0.471	0.595	0.080
	ANFIS (2)	Gbell	Linear	5,6	32	Hybrid	0.552	0.619	0.017
	ANFIS (3)	Gbell	Linear	5,6	32	Hybrid	0.363	0.596	0.086
AE-N	ANFIS (4)	Gbell	Linear	5,6	32	Hybrid	0.262	0.535	0.114
	ANFIS (1)	Gbell	Linear	5,6	32	Hybrid	0.203	0.606	0.103
	ANFIS (2)	Gbell	Linear	5,6	32	Hybrid	0.217	0.651	0.078
	ANFIS (3)	Gbell	Linear	5,6	32	Hybrid	0.531	0.582	0.073
AE-P	ANFIS (4)	Gbell	Linear	5,6	32	Hybrid	0.534	0.696	0.033
	ANFIS (1)	Gbell	Linear	5,6	32	Hybrid	0.559	0.574	0.030
	ANFIS (2)	Gbell	Linear	5,6	32	Hybrid	0.363	0.555	0.115
	ANFIS (3)	Gbell	Linear	5,6	32	Hybrid	0.599	0.627	0.029
HT-A	ANFIS (4)	Gbell	Linear	5,6	32	Hybrid	0.556	0.662	0.089

	ANFIS (1)	Gbell	Linear	5,6	32	Hybrid	0.271	0.657	0.028
	ANFIS (2)	Gbell	Linear	5,6	32	Hybrid	0.289	0.660	0.025
	ANFIS (3)	Gbell	Linear	5,6	32	Hybrid	0.435	0.631	0.085
HT-W	ANFIS (4)	Gbell	Linear	5,6	32	Hybrid	0.404	0.643	0.050

4. Conclusion

In terms of life cycle assessment, these methods were combined and applied in studies aimed at agriculture, climate and engineering studies. The field of application is characterized by algorithms capable of learning from examples and reproducing. Its foundations are based on intensive data analysis and its application in LCA can be seen in solving environmental problems. Being considered efficient and economical substitutes for conventional procedures, these artificial intelligence approaches provide a high level of capability, deal with the complexities of uncertain, interactive and dynamic problems and contribute to the minimization of uncertainties associated with the process.

5. Acknowledgments

The authors are grateful to the Coordination for the Improvement of Higher Education Personnel (CAPES)- Finance Code 001 and Industrial Engineering Graduate Program (PEI) at the Federal University of Bahia (UFBA)for supporting this research.

6. References

[1] Azapagic, A.; Millington, A.; Collett. A. 2006. A methodology for integrating sustainability considerations into process design. Chemical Engineering Research and Design. 84(A6): 439–452.

[2] Froemelt, A., Dürrenmatt, D. J., Hellweg, S. 2018. Using Data Mining To Assess Environmental Impacts of Household Consumption Behaviors. Environmental Science & Technology. 52 (15) 8467-8478.

[3] Hosseinzadeh-Bandbafha H., Nabavi-Pelesaraei A., Majid K; Ghahderijani M., Chau, K. 2018. Application of data envelopment analysis approach for optimization of energy use and reduction of greenhouse gas emission in peanut production of Iran.Journal ofCleaner Production. 172. 1327–1335.

[4] Hosseinzadeh-Bandbafha H., Safarzadeh D., Ahmadi, E. 2019. Combined life cycle assessment and artificial intelligence for prediction ofoutput energy and environmental impacts of sugarcane production. Science of the Total Environment. 664.1005-1019.

[5] Hosseinzadeh-Bandbafha, H., Chau, K. W. 2018. Integration of artificial intelligence methods and life cycle assessment to predict energy output and environmental impacts of paddy production. Science of the Total Environment. 631–632; 1279–1294.

[6] Hosseinzadeh-Bandbafha, H., Safarzadeh, D., Ahmadi E, Nabavi-Pelesaraei, A., Hosseinzadeh-Bandbafha, E. 2016. Applying data envelopment analysis to evaluation of energy efficiency and decreasing of greenhouse gas emissions of fattening farms. Energy.1-11.

[7] Hou, P., Jolliet O., Zhu, J., Xu, M. 2020. Estimate ecotoxicity characterization factors for chemicals in life cycle assessment using machine learning model. Environment International. 135.

[8] Huntingford, C., Jeffers, E. S., Bonsall, M. B, MChristensen, H., Lees, T., Paramesh, V., Arunachalam, V., Nikkhah, A., Das, B., Yang, H. 2019.Machine learning and artificial intelligence to aid climate change research and preparedness. Environ. Res. Lett. 14.

[9] ISO (INTERNATIONAL ORGANIZATION FOR STANDARDIZATION). 2006 14044: Environmental managements - lifecycleassessments – requirementsandguidelines. InternationalOrganization for Standardization.

[10] Kaab, A., Sharifi, M., Mobli, H., Nabavi-Pelesaraei, A., Chau, K. W. 2019. Use of optimization techniques for energy use efficiency and environmental life cycle assessment modification in sugarcane production. Energy.181. 1298-1320.

IOP Conf. Series: Materials Science and Engineering 1250 (2022) 012003 doi:10.1088/1757-899X/1250/1/012003

Mechanical Properties and Water Absorption of Glass Fiber Reinforced Composite Laminates made by Varying Stacking Sequence

PAGLICAWAN, Marissa A.[1a], EMOLAGA, Carlo S.[1], CUSTODIO, Cyron L.[1], SUDAYON, Johanna Marie B.[1], DE LARA, Ryan John L.[2]

[1] Department of Science and Technology, Industrial Technology Development Institute, General Santos Avenue, Bicutan, Taguig City, Philippines 1631

[2]Wesleyan University of the Philippines, Mabini Extension, Cabanatuan City, Nueva Ecija, Philippines 3100

[a]mapaglicawan@itdi.dost.gov.ph

Abstract. This study aims to demonstrate the effects of various stacking sequences of different glass fiber architecture forms on the mechanical properties of glass fibre reinforced polyester composites. The composites were manufactured by hand lay-up technique and its mechanical properties were characterized in terms of tensile, flexural, and drop-ball impact tests as well as water absorption. The mechanical properties of the glass fiber reinforced polyester were somewhat affected depending on the stacking sequences from single alternating order, double alternating order, and triple alternating order of chopped strand mat and plain-woven fabric piled in ten (10) to fourteen (14) layers. The laminates consisting of alternating sequence manner of plain-woven fabric and chopped strand mat stacked in different layers have no significant effect on the mechanical properties. Stacking sequence of double alternating layer piled in twelve (12) layers was found to have better tensile strength than single and double alternating order of two different glass fiber form while the triple alternating manner piled in twelve (12) layers resulted to higher tensile modulus. The double alternating order of different glass fiber forms stacked in twelve (12) layers resulted to an improvement in flexural properties. Some selected laminates exhibited no delamination in drop-ball impact test and minimal water absorption when immersed in water for 30 days for all the laminates.

1. Introduction

Glass fiber is a common and versatile synthetic fiber in reinforcing thermosets and thermoplastics. They exhibit useful bulk properties such as hardness, transparency, resistance to chemical attack, stability, and inertness as well as desirable fiber properties such as strength, flexibility, and stiffness [1,2]. The application of this material has been used in different industries such as construction and automotive industries. However, there are few main factors that may affect the mechanical performance of the glass fiber reinforced polymer composites such as fiber quality, fiber dispersion, orientation, matrix quality, interfacial strength, porosity, and manufacturing process [3]. As to improve the strength of a composite material, one should be able to design and form its internal and external structures advantageous to certain applications [4]. The fiber reinforcement quality and polymer matrix quality are the most important factors to be considered in designing a composite material. Polyester provides acceptable strength as binding reagent for glass fibers and is used for general industry purposes such as part fabrication and low-cost molds. A study about fiber reinforced laminated composites has been done evaluating the influence of stacking sequence on the mechanical properties using carbon/glass fiber and epoxy [5]. Another study has reported that by reinforcing glass fiber with polyester with increasing the specimen size geometry has led to an increase in energy at break, peak load, critical strain energy release rate and critical stress [6]. Most of these recent research studies were also focused on the mechanical and impact properties of the continuous plain-woven glass fiber reinforced polymer composites [7-10] and investigation of the mechanical and impact properties of short glass fiber reinforced [11-12]. Researchers also studied the notch sensitivity of short and 2D plain woven glass fibers polymer composites [13]. Moreover, the effect glass fiber stacking schedule on the mechanical properties of glass fiber reinforced polyester composite was studied only using five layers mainly composed of plain woven, short fiber, and sandwich layer glass composites [14], and different stacking sequences of glass fiber-epoxy composites [15] and in different natural fiber and synthetic fibers [16]. Therefore, the objective of this study is to investigate the influence of various stacking sequences consisting of different glass fiber architectures on the mechanical properties and water absorption glass fiber polyester composites.

2. Experimental Procedures

2.1. Materials

The glass fibers with architecture of fiber orientations such as chopped strand mat (randomly oriented fibers) and plain-woven fabric (wrap and fill yarn at 0° and 90° orientation were used in this study. Chopped stand mat and plain-woven fabric densities are 450gsm, 600gsm and 800gsm respectively. The polymer matrix used is a polyester resin in isophthalic and orthophthalic grades for outer/exposed and inner/unexposed part respectively. Styrene monomer with 10wt% was used to lessen the viscosity of the resin, and cobalt accelerator with 0.5wt% and methyl ethyl ketone peroxide or MEKP with 1wt% were used to increase the catalytic activity.

2.2. Fabrication of polymer composites

Glass fiber reinforced unsaturated polyester composites were fabricated using traditional hand lay-up method. Through this technique, each reinforcement layer was placed in a mold and wet with resin using brush and roller. The use of brush and roller was done manually in order to disperse the resin throughout the layers of fibers. The labels and stacking sequence of each composite were shown in Table 1. Laminates were then cut accordingly to the standards of characterization.

Table 1. Lamination structure of fabricated composite samples.

Sample Label	Lamination Schedule
PWR600	10 layers of plain-woven fabric $[(WR600)_{10}]_1$
PCM450	10 layers of chopped strand mat $[(CSM450)_{10}]_1$
7CWSA	7 layers of single alternating $[(CSM450)_1(WR600)_1]_3[(CSM450)_1]_1$
8CWSA	8 layers of single alternating $[(CSM450)_1(WR600)_1]_3[(CSM450)_1(CSM300)_1]_1$
10CWSA	10 layers of single alternating $[(CSM450)_1(WR600)_1]_5$
10CWDA	10 layers of double alternating $[(CSM450)_2(WR600)_2]_5$
10CWTA	10 layers of triple alternating $[(CSM450)_3\ (WR600)_4(CSM450)_3]_1$
10WCTA	10 layers of triple alternating $[(WR600)_3\ (CSM450)_4(WR600)_3]_1$
12 CW600DA	12 layers of double alternating $[(CSM450)_2\ (WR600)_2]_6$
14 CW600DA	12 layers of double alternating $[(CSM450)_2(WR600)_2]_7$
12CW800DA	12 layers of double alternating $[(CSM450)_2(WR800)_2]_6$
14 CW800DA	14 layers of double alternatin $[(CSM450)_2(WR800)_2]_7$

2.3. Characterizations

The tensile tests after conditioning at 25 ◦C were carried out on a Universal Testing Machine (UTM) model Instron 5585H (Instron, Norwood, MA, USA) equipped with a 10 kN capacity load cell. The tests were performed in accordance with ASTM standards D3039. The crosshead speed used was 5 mm/min for tensile testing. The flexural test procedures and values were obtained in accordance with ASTM D790 and ISO 178 using Shimadzu Universal Testing Machine (AGS-50kNXD, Shimadzu, Tokyo, Japan) with specimens of nominal dimensions of 50 mm × 25 mm × 2.5 mm, a span length of 90 mm, and a crosshead speed of 0.71 mm/min. The average values of the mechanical properties were obtained from 6 specimens. The drop-ball impact test was performed using different working conditions wherein the weight is 1.5 kg and 1m drop height. Condition A was suspended in sand and B was suspended in air with impact energy of 14.73kg.m^2/s^2. Condition C was suspended in sand and D was in air with impact energy of 29.43kg.m^2/s. The water absorption test was done in accordance with ASTM D5229. The specimens were cut into 25.5 x 25.5 x2 mm^3. Prior to absorption experiments, five specimens were dried in an oven for 25 hrs at 102±3°C. The values of water absorption in percentage were calculated as follows: $WA\ (\%)_t = (Mc - Mo)/Mo \times 100$, where WA (t) is the water absorption at time t, *Mo* is the mass of the dried specimen and *Mc* is the mass of the specimen as a function of immersion time.

3. Results and Discussion

Figure 1 shows the tensile properties of composite laminates consisting of different stacking sequences of WR and CSM. Composites of plain-woven fabric and chopped strand mat were also fabricated as control and comparison. As expected, the tensile strength of WR600 is higher than CSM450 due to their different architecture forms. Although this was not the case in the tensile modulus. When these glass fibers in different architecture forms were combined, the results were dependent on different stacking sequences and number of layers. For instance, comparing the laminates coded 7CWSA, 8CWSA, and 10CWSA, where the stacking sequence was arranged in single alternating manner of CSM followed by WR, the tensile strength and tensile modulus have not much significant difference. The results showed that the tensile strength was 193MPa, 202MPa and 197MPa while the tensile modulus was 11.2GPa, 11.9GPa and 10.2GPa respectively. If the stacking sequence of WR and CSM was double alternating manner as in the case of 10CWDA, it was observed that no significant difference of the tensile properties. On the other hand, when the glass fibers were stacked in triple alternating of 10CWTA wherein the WR was sandwiched between the three layers of CSM. The tensile strength was reduced to 169MPa. However when the three layers WR600 was placed as part of the skin of the laminates and the CSM450 was sandwiched by the three layers of WR600 as in the case of laminate coded 10WCTA, a significant increase in tensile strength (231 MPa) was observed. Both the tensile modulus exhibited similar results. Increasing the number of layers from 10 to 12 in double alternating manner exhibited slight increase in tensile strength of from 197MPa to 209MPa but decreases when the layers was increased to 14 (168MPa). There was increase in tensile modulus from 11.3GPa to 19.7GPa and 15.7GPa. Replacing WR600 to WR800, laminates coded 12 CW800DA also showed a better tensile strength of 241MPa which is the highest so far in all combinations of different architecture forms of glass fiber. However further increase in the number layers to 14 layers such as in 14CW800DA, the tensile strength was decreased. The tensile modulus of the two laminates were 18.2GPa and 15.37GPa respectively which is almost similar to 12CW600DA and 14CW600DA.

The behavior of flexural properties of these laminates was quite different from the tensile properties (Fig. 2). The flexural strength of 10 layers of pure WR600 resulted to 289MPa and tensile modulus of 17.7GPa. These results were not high as compared to 10 layers of pure CSM450 having 248MPa flexural strength. However, the flexural modulus of WR600 is higher than CSM450 (17.7GPa and 9.71GPa respectively). The laminates coded 7CWSA, 8CWSA and 10CWSA showed almost similar value of flexural strength of 253MPa, 262MPa and 254MPa respectively and flexural modulus of 8.3GPa, 8.83GPa, and 8.82GPa respectively. These laminates are made of alternating CSM and WR with increasing the number of layers. Despite the increasing number of layers, the results of flexural properties did not matter at all. Comparing the 10 layers of CWSA, a single alternating and 10 layers of CWDA, double alternating manner, no significant difference of flexural strength and not much improvement of flexural modulus was observed. The flexural strength and flexural modulus were further decreased if the stacking sequence was changed to triple alternating of CSM and WR such as in the case of laminate coded 10CWTA. However good results were obtained when the three layers of CSM450 were sandwiched with 3 layers of WR600 as in the case of laminate coded 10WCTA. The flexural properties of 236MPa and 14.9GPa. Increasing the number of layers from 10 to 12 in laminate coded 12CW600DA of double alternating resulted to higher flexural strength of 385MPa and flexural modulus of 12.7GPa. However further increasing the number of layers specifically 14 layers (laminate coded 14CW600DA), it did not improve the flexural properties. On the other hand, replacing the WR600 to WR800 of the same layers (12CW800DA), an improvement of flexural strength of 444MPa was obtained. But decreases again with further increase of 2 layers (300MPa). Both of the flexural modulus did not give high result.

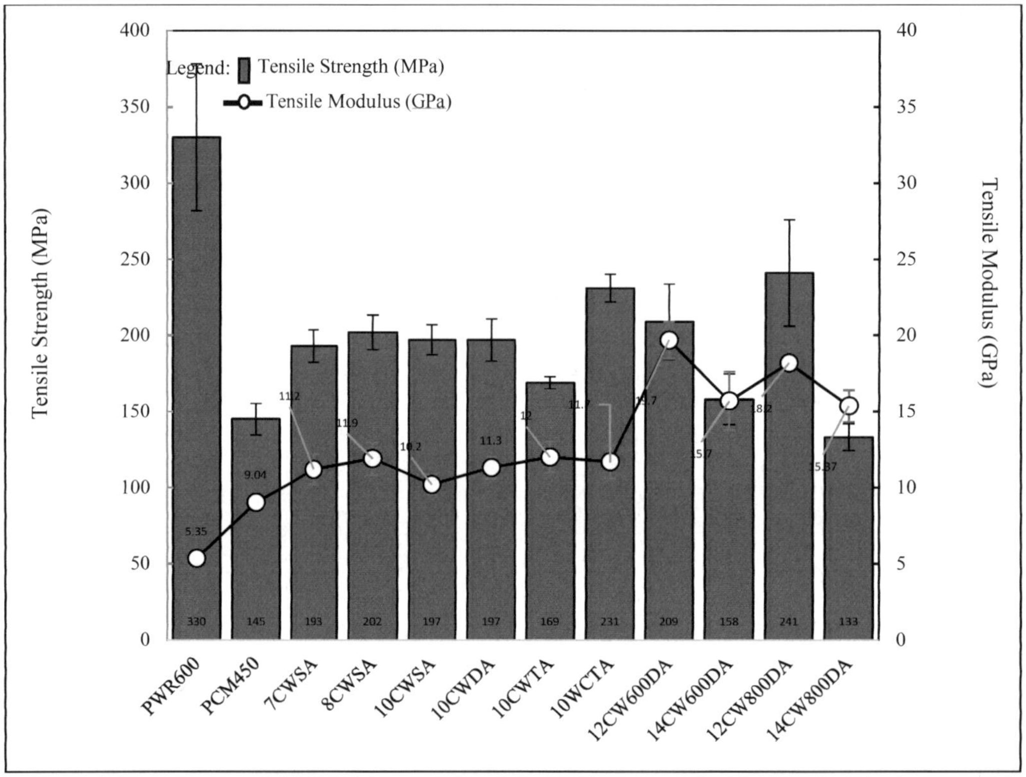

Figure 1. Tensile properties of different stacking sequence

Drop-ball impact of the laminates was performed of some selected laminates to determine how the impact energy propagates through the samples that may cause cracking and delamination. Figure 3a shows that no delamination occurred in laminates coded 10CWDA in different testing condition. Similarly, the laminates coded 14CW600DA (Fig.3b-d) with different dimensions to find out the effect on the dimension of the test specimens, which resulted to no delamination.

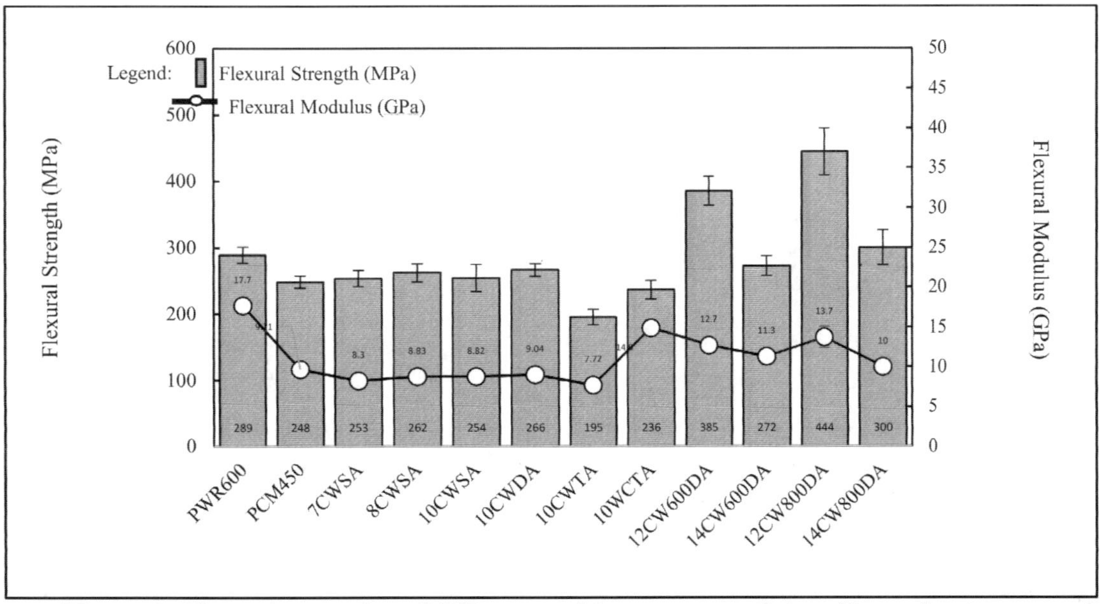

Figure 2. Flexural properties of different stacking sequence of glass fiber polyester composites

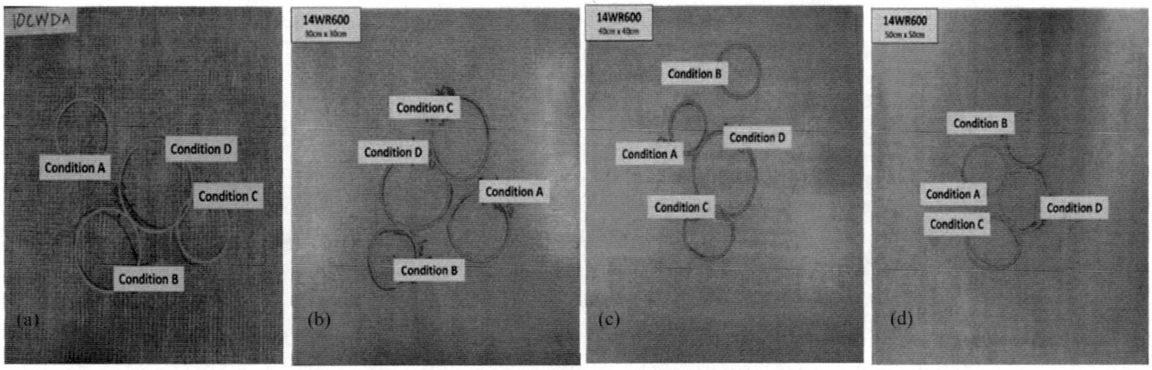

Figure 3. Drop ball impact behavior of (a) 10CWDA (b) 14CW600DA with specimen size of (a) 30cm x 30cm (b) 40cm x 40cm (c) 50cm x 50cm

Figure 4 shows the water absorption of the composite as a function of time. It can be seen that the water absorption for all the laminates using different stacking sequence slight increases as the time of immersion increases up to 16 days (400hrs) and saturation point reached up to 30 days. The value of water absorption for 8hrs which ranges from 0.08% to 0.41% which is not dependent on the number layers for WR600 or WR800 and CSM. This behavior was also observed in composite with combination of different fibers architecture form. After one-month immersion, all the composites exhibited minimal absorption ranges from 0.41% to 0.62%.

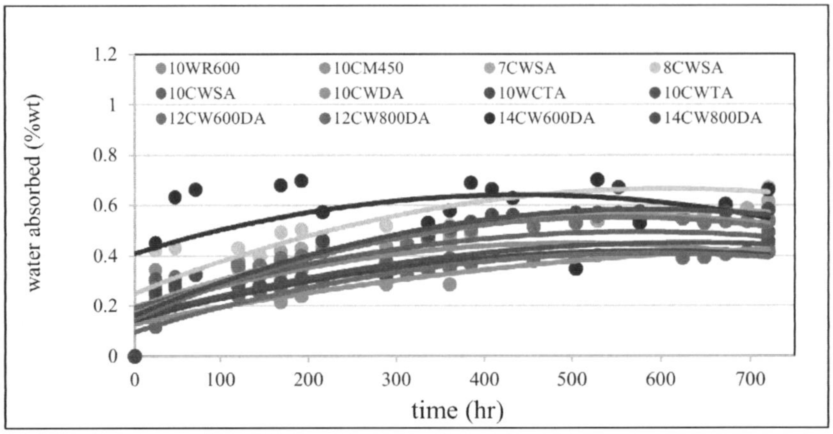

Figure 4. Water absorption of different stacking sequences laminates for 30 days

4. Conclusions

The laminates coded 10WR600 which consists of plain-woven fabric resulted to highest tensile strength. Among the different stacking sequences using different architecture glass fiber form, 12CW800DA having this stacking sequence of $[(CSM450)_2(WR800)_2]_6$ resulted to highest tensile strength (241 MPa) and flexural strength (444 MPa) then followed by the laminates coded 10WCTA having this stacking sequence of $[(CSM450)_3\ (WR600)_4(CSM450)_3]_1$. Similarly, these laminates also gave one of the highest modulus. Based on the drop-ball impact energy and different testing conditions, no extensive delamination was seen. The stacking sequences of different glass fiber architectures have no significance effect on water absorption of the laminates.

References

[1] Wallenberger FT 1999 Structural Silicate and Silica Glass Fibers. In Advanced Inorganic Fiber Processes, Structures, Properties, Applications; Wallenberger, FT Ed.; Kluwer Academic Publishers: New York NY USA 129.

[2] Wallenberger FT, Watson JC, Li H 2001 Glass Fibers @2001 ASM International. In ASM Handbook Volume 21 Composites ASM International Materials Park Cleveland OH USA

[3] Joy J, Jaroszewski M, Praveen KM, Thomas S and Haghi R 2018 Engineering Technologies for Renewable and Recyclable Materials: Physical-Chemical Properties and Functional Aspects Innovations in Physical Chemistry 1st ed Apple Academic Press Oakville, Canada

[4] Ashton JE, Halphin JC and Petit PH 1998 On Composite Materials, Technonomic Publishing Co Inc, Stanford Comm 13 53.

[5] Jeshti DK, Mandal P, Nayak RK and Rout AK 2018 *Procedia Manufacturing* 20 530.

[6] Azizah H and Mohd R 2011 *J. of Appl. Polym Sci* 5 62.

[7] Velmurugan R, Solaimurugan S 2007 *Composites Science and Technology* 67 61.

[8] Esnaola A, Tena I, Aurrekoetxea J, Gallego I, Ulacia I 2016 *Composites Part B: Engineering* 85 1.

[9] Kanerva M, Sarlin E, Hoikkanen M, Rämö K, Saarela O, Vuorinen J *2015 International Journal of Adhesion and Adhesives* 59 40.

[10] Bergmann T, Heimbs S, Maier M 2015 *Composite Structures* 125 362.

[11] Faudree MC, Nishi Y 2010 *Materials Transactions* 51 2304.

[12] Dhakal HN, Zhang ZY, Richardson MOW, Errajhi OAZ 2007 *Composite Structures*81 559.

[13] Elbadry EA, Abdalla GA, Aboraia M and Oralby EA 2017 *J. Reinf Plastics & Comp. 36 1092.*

[14] Elbadry EA, Abdalla GA, Aboraia M and Oralby EA 2018 *J. Material Sci Eng* 7 1.

[15] Bere P, Popescu A, Dudescu C, Hancu L 2017 MATEC Web of Conferences 112.

[16] Das SC, Paul D, Grammatikos SA, Siddiquee Md AB, Papatzani S, Koralli P, Islam J MM, Khan MA, Shauddin SM, Khan RA, Vidakis N, Petousis M 2021 *Composites Structure* 276 114525

IOP Conf. Series: Materials Science and Engineering 1250 (2022) 012004 doi:10.1088/1757-899X/1250/1/012004

Influence of Abaca Fiber on the Performance of Abaca-Glass Fiber Reinforced Polymer Composites

PAGLICAWAN, Marissa A[1a], EMOLAGA, Carlo S[1], SUDAYON, Johanna Marie B.[1] and CUSTODIO, Cyron L[1].

[1]Department of Science and Technology, Industrial Technology Development Institute, Bicutan, Taguig City 1631, Philippines

[a]mapaglicawan@itdi.dost.gov.ph

Abstract. Woven abaca fiber also known as "Sinamay" was usually used in textile and handcraft industries. In this paper, the use of woven abaca fiber was utilized as a potential component in a fiber reinforced polymer (FRP) composite material. It also aims to study the effects of increasing abaca fiber layer on the mechanical and water absorption properties of the composite. The composite laminates having 1, 3, 5, and 7 layers of woven abaca stacked in between of glass fiber mats: plain-woven fabric and chopped strand and polyester resin as binder were fabricated by hand lay-up technique. These abaca-glass fiber reinforced polyester composite specimens were characterized using tensile test, flexural test, izod impact test, and water absorption test. Upon introducing and increasing the amount of abaca layers in the composite, the results showed that its properties were significantly changed. The mechanical strength was decreased while water absorbed was increased as the amount of abaca layer increases.

1. Introduction

Abaca fiber also known as “Manila Hemp” which is obtained from pseudo-stem of banana family (Musa Textilis Nee) exhibits exceptional properties sufficient for modern environmental-friendly applications in handicraft and industrial products. Philippines is the largest producer of abaca fiber supplying about 90% of the world's requirement for the production of cordage, specialty papers (for currency note, stencil paper, teabag, coffee filter/cup, capacitor, and insulation paper, etc.), textiles, furniture, handicrafts, novelty items, meat casing, and composites or automotive and construction and industrial products [1]. According to studies, abaca fibers possess better mechanical properties than other natural fibers and can be compared to synthetic fibers. Abaca fiber has great potential for different applications due to its high tensile strength, flexural strength, and salt degradation resistance [2-4]. In addition to this, hybridization of natural fibers with synthetic fibers was also found to improve the mechanical properties of the materials as a composite material [5]. Introducing abaca fibers in composites could also improve the biodegradability of a material. The capabilities of abaca fiber as promising alternative to synthetic fibers used in composite materials has attracted several industries. Nowadays, abaca fibers were used in automotive, construction, and marine industries. DaimlerChrysler’ researchers, along with the manufacturing specialists, Rieter Automotive and supplier specialists, Manila Cordage was commended on 2005 because of their innovation on using abaca fiber reinforced polypropylene composites on their Mercedes Benz’s exterior of a passenger car as a covering for the spare wheel recess [6]. On the other hand, another application of abaca fiber reinforced epoxy composite as retrofitting materials were investigated and applied wherein showed that there is a significant increase of compressive strengths with the introduction of epoxy and abaca fiber as a confining material [7]. However, studies also show some drawbacks of using natural fiber as reinforcement for composite. Its poor compatibility of fibers and matrix, the degrading behavior of the composites when exposed to various weather conditions, and its hydrophilic nature were few of the challenges of using natural fibers [8]. Abovementioned studies have led to conduct an experiment which specifically demonstrates the effect of using abaca fiber layer in form of woven, increasing the amount of the fiber layer loading, and using polyester resin as binder in this natural/synthetic fiber reinforced polymer composite material on its mechanical properties and water absorption.

2. Experimental Procedure

2.1 Materials

S-2 (Streaky Two) grade abaca fibers in the form of plain fabrics in the $0^0/90^0$ direction containing 19 fabrics in warp and 11 fabrics in the welf direction were used in this study, similar to that in another paper [9], which was taken from Bicol Region, Philippines. The glass fibers used in the study are in the form of plain-woven fabric ($0^0/90^0$) (WR600), 600g/m^2 and chopped strand mat (CSM300) 300g/m^2. The matrix used was orthophthalic polyester, with cobalt naphthenate as the accelerator and methyl ethyl ketone (MEKP) as the hardener. These chemicals were purchased from Polymer Products Inc., Philippines.

2.2 Fabrication of specimens

Abaca-glass fiber reinforced unsaturated polyester composites were fabricated using traditional hand lay-up method. Through this technique, each reinforcement layer was placed in a mold and wet with resin using brush and roller. The use of brush and roller was done manually in order to disperse the resin throughout the layers of fibers. The labels and stacking sequence of each composite were shown in Table 1. The resin consists of 0.5wt% cobalt naphthenate/accelerator, 10wt% styrene and 1 wt% MEKP. Abaca-glass fiber reinforced polyester composite laminate was cured under room temperature and cut accordingly to the standards of characterization tests.

Table 1. Lamination structure of fabricated composite specimens.

Specimen Label	Lamination Schedule
HSW6C3	(1) Surface Mat, (1) WR600, (1) CSM300
HSW6A1*C3	(1) Surface Mat, (1) WR600, (1) Abaca, (1) CSM 300
HSW6A3*C3	(1) Surface Mat, (1) WR600, (3) Abaca, (1) CSM 300
HSW6A5*C3	(1) Surface Mat, (1) WR600, (5) Abaca, (1) CSM 300
HSW6A7*C3	(1) Surface Mat, (1) WR600, (7) Abaca, (1) CSM 300

2.3 Characterizations

2.3.1 *Mechanical properties.* The tensile tests after conditioning at 25 ◦C were carried out on a Universal Testing Machine (UTM) model Instron 5585H (Instron, Norwood, MA, USA) equipped with a 10 kN capacity load cell. The tests were performed in accordance with ASTM standards D3039. The crosshead speed used was 5 mm/min for tensile testing. The flexural test procedures and values were obtained in accordance with ASTM D790 and ISO 178 using Shimadzu Universal Testing Machine (AGS-50kNXD, Shimadzu, Tokyo, Japan) with specimens of nominal dimensions of 50 mm × 25 mm × 2.5 mm, a span length of 90 mm, and a crosshead speed of 0.71 mm/min. The specimens were prepared according to ASTM D256. Impact test was done on an impact pendulum tester to know if how much impact energy that the composite material can resist upon impact. Values were acquired in each specimen and were then evaluatedThe average values of the mechanical properties were obtained from 6 specimens.

2.3.2 Water absorption test. Water absorption test was done to determine the ability of composite material to absorb moisture or water. Standard procedure was followed according to ASTM D570. The specimens were thoroughly dried at normal conditions and stored in a desiccator after fabrication. Note that for each measurement, specimens were removed from the water and the surface water was wiped off using blotting paper. Weight of the specimens before and after submersion were recorded each time. Water absorbed is expressed as increase in weight percent:

$$\text{Water Absorbed } (\%\text{wt}) = \left(\frac{W_t - W_i}{W_i}\right) \cdot 100 \qquad (1)$$

where: W_i = initial dried weight

W_t = weight of the specimen as a function of immersed time

The test has undergone for 720 hours (30 days). The lower the water absorbed the greater its water resistance.

3. Results and Discussion

The results showed that with the addition of abaca, there has been a significant change on the tensile, flexural, and impact strength of the material as well as its water absorption properties. Tensile behaviour of the composites with increasing amount of woven abaca fiber layers were shown in Figure 1. This demonstrates that from the specimen HSW6C3 (composed of one layer of plain-woven fabric and chopped strand mat, tensile strength was decreased with introduction of one layer of woven abaca in between two glass fibers as in the case of specimen HSW6A1*C3. The tensile strength was decreased from 160MPa to 97.5MPa. Also, continuous addition of woven abaca fiber layer in the composite, decreases the tensile strength and modulus of the material. This behavior was also observed in our previous studies [10].

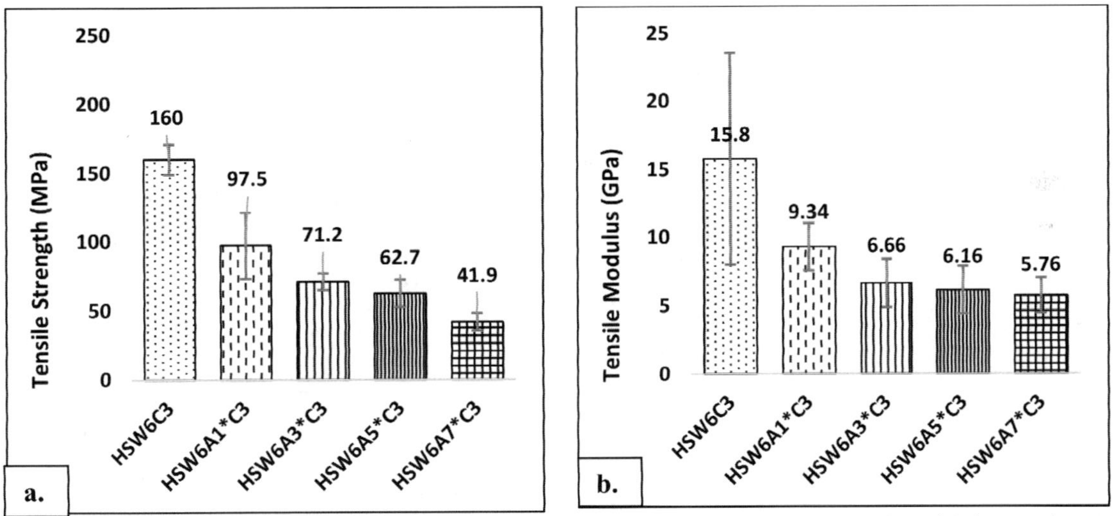

Figure 1. Tensile properties of abaca-glass fiber reinforced polyester composite with increasing layers of woven abaca. (a) Tensile Strength, (b) Tensile Modulus.

On the other hand, introduction of woven abaca fiber layer to the composite shows different flexural and impact behaviors as presented in Figure 2 and 3. Flexural strength was improved from 148MPa to 150MPa upon adding one layer. In the same manner, impact strength was also improved from 646MPa to 485MPa. Due to this results, addition of abaca fiber was optimized up to 7 layers in order to observe the changes. Though the flexural strength and impact strength were improved upon the addition of one layer of abaca, further addition of more than one layer, has decreased it flexural and impact strengths. It seems that the maximum amount of abaca fiber layer that can be added to produce favorable mechanical properties of hybrid abaca-glass fiber composite is only limited at some extent. In this case, HSWA1*C3 with one layer of woven abaca stacked in between chopped strand mat and woven roving mat was the optimum loading. It could be because of the adhesion of abaca to resin to abaca fiber is weak and could not sustain the bending loading applied [9].

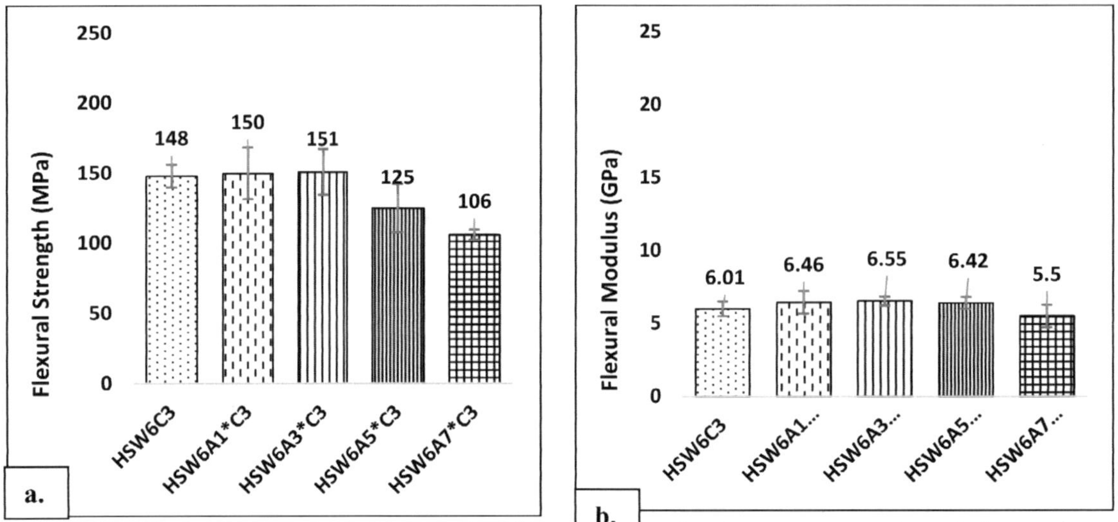

Figure 2. Flexural properties of abaca-glass fiber reinforced polyester composite with increasing layers of woven abaca. (a) Flexural Strength, (b) Flexural Modulus.

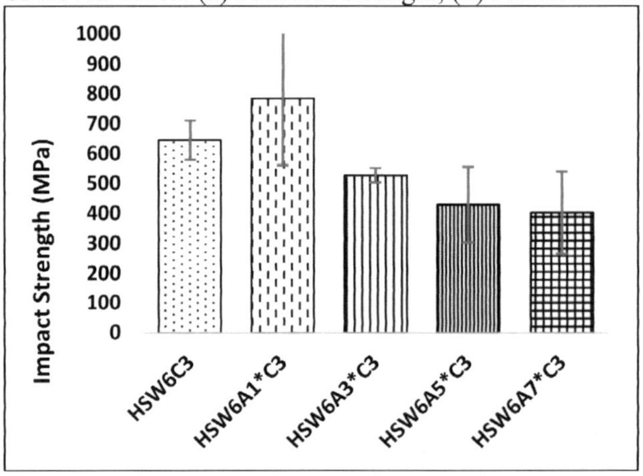

Figure 3. Impact properties of abaca-glass fiber reinforced polyester composite with increasing layers of woven abaca

Water absorption test (Figure 4a) shows the behavior of the lamination specimens upon submersion in water within 720 hours. This graphical curve illustrates the water absorbed as a function of time and shows that specimens HSW6C3 and HSW6A3*C3 have reached their saturation points at around 400hrs to 500hrs. Moreover, in Figure 4b, with the increasing amount of abaca fiber layer, the amount of water that each lamination specimen also increases. This suggests that the hydrophilic nature of abaca fiber affected the water absorption of the abaca-glass fiber reinforced polyester composite.

.

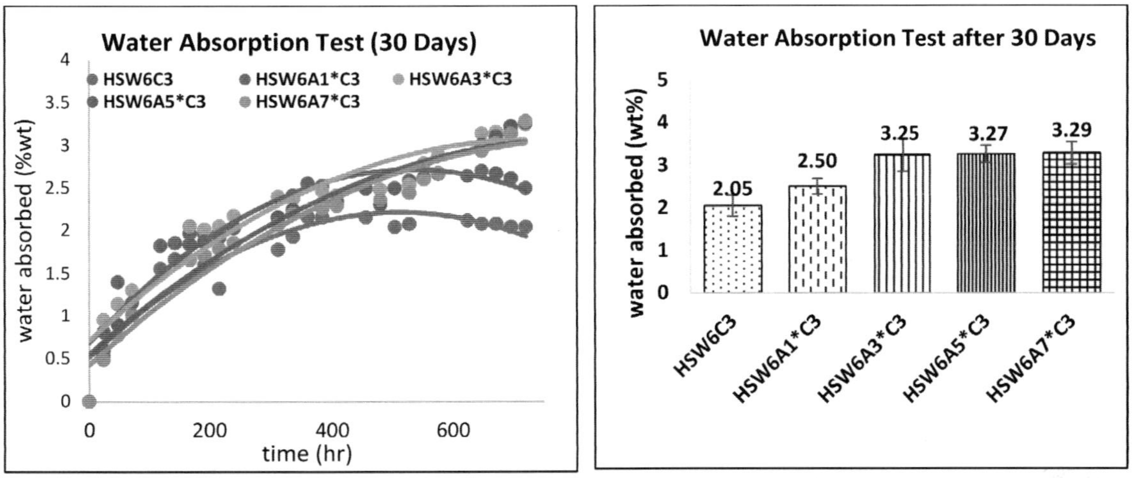

Figure 4. (a) Water absorption curve (b) water absorption results after 30 days of submersion, of abaca-glass fiber reinforced polyester composite with increasing layers of woven abaca

4. Conclusion

Introduction of woven abaca fiber as one of the reinforcements and using polyester resin as polymer matrix in a composite material was successfully done in this study. The loading of abaca fiber has affected the mechanical and water absorption properties of the composite material. The tensile strength and modulus decrease as the woven abaca fiber loading increases. Moreover, abaca-glass fiber reinforced composite was optimized at one abaca layer as a result of flexural and impact tests. Water absorption proves that the hydrophilic nature of abaca fiber showing that with increasing amount of abaca fiber layer, the amount of water absorbed increases. Further studies on stacking sequence like alternating layers of abaca and glass fibers is recommended.

References

[1] Philippine Fiber Industry Development Authority 2015 *Philippine Abaca helps in global environment conservation* (Hibla Newsletter) vol 1 no 1

[2] Galande V and Chaudhari D 2017 International Conference on Ideas, Impacts and Innovation in Mechanical Engineering 5 6

[3] Huang G 2005 vol 30 (Materials and Design) pp 3931-3934

[4] Moreno L O, Parac A A, Ocon F L, Gapasin R M 2010 *Philippine Journal of Crop Science* 35 1 113

[5] Ramnath V, Kokan S, Raja R, Sathyanarayanan R, Elanchezhian C, Prasad A, Manickavasagam V 2013 Materials & Design 357–366

[6] *DaimlerChrysler Awarded for Banana Fibre Use in Mercedes A Class* 2005 (CompositesWorld, NetComposites)

[7] Dancel R 2018 *Abaca Fiber as a Retrofitting Material* (Research fora International Conference)

[8] Vinayagamoorthy R 2017 *Journal of Reinforced Plastics and Composites* 36 21 1577.

[9] Paglicawan MA, Rodriquez MP, Celorico JR 2020 *Polym. Compos.* 411 763.

[10] Paglicawan MA, Emolaga CS, Sudayon JMB, Tria KB 2022 *Polymers* 13 2719.

[11] Sathish P, Kesavan R, Ramnath V 2015 *ARPN Journal of Engineering and Applied Sciences* 10 5534

IOP Conf. Series: Materials Science and Engineering 1250 (2022) 012005 doi:10.1088/1757-899X/1250/1/012005

Controllable fabrication of silver nanofilament for using as tip in tip-enhanced Raman spectroscopy

O Phonrachom[1, 2], P Pienpinijtham[2, 3], C Sapcharoenkun[4]

[1]Petrochemistry and Polymer Science, Chulalongkorn University, Bangkok, 10330 Thailand.
[2]National Nanotechnology Center of Advanced Structural and Functional Nanomaterials, Faculty of Science, Chulalongkorn University, Bangkok, 10330 Thailand.
[3]Department of Chemistry, Faculty of Science, Chulalongkorn University, Bangkok, 10330 Thailand.
[4]National Nanotechnology Center, NSTDA, Pathum Thani, 12120 Thailand.

Abstract. Tip-enhanced Raman spectroscopy (TERS) is a technique offering chemical information at a single molecule level as well as chemical fingerprints with nanoscale spatial resolution. Currently, TERS tip fabrication in terms of size, shape, and stability has still been challenged due to its low reproducibility. In this work, electron beam irradiation and chemical bath deposition (CBD) were used to create a silver nanofilament (AgNF) on an atomic force microscope (AFM) tip, and it was further used as a TERS tip. Conditions, *i.e.,* accelerated voltage and irradiation time, in the electron beam irradiation with different Ag thin film thicknesses in the production of the AgNF tip were investigated. The highest density of AgNF of the TERS probe was achieved with 52.5×10^9 NF/cm^2 by using an accelerated voltage of 20 kV and an irradiation time of 20 minutes. This study will pave the way for the use of the AgNF tip as the TERS tip.

1. Introduction

Tip-enhanced Raman spectroscopy (TERS) is a technique that combines nanoscale scanning probe microscopy (SPM) with Raman spectroscopy, which provides chemical information on a single molecule with high spatial resolution. TERS has been used in many fields, such as material and biological specimen identification. The TERS probe has been created to improve the spatial resolution of the images obtained from Raman imaging. The tip apex in TERS is sharp, affecting highly localized surface charge density and electric field intensity. Then, the production of the tip to create a sharp AgNF tip with controllable thickness and shape is important to fabricating the TERS tip. Metal nanoparticles are widely used as material for probes in TERS, due to their unique electrical and optical properties, Silver (Ag) is one of the metals that exhibits good thermal and electrical conductivity under ambient conditions [1-3].

Nowadays, there are several methods to fabricate AgNF tips. For example, thermal evaporation is an uncontrollable technique for tip apex fabrication because of the random deposition [4-5]. Ion beam milling gives nanoscale and high reproducibility, but it requires professional operation [6-7]. Electrochemical etching produces a smooth surface and a controllable tip apex size, but this technique is complicated in preparation. Cuncheng *et al.* reported the fabrication of Ag composites and Ag nanostructures with a Ag composites precursor. The electron beam irradiation (EBI) with low or medium current density created Ag composites, while the high current density created small AgNPs [8].There are two steps for AgNF tip fabrication. Firstly, chemical bath deposition (CBD) is a method for producing functional oxide films by immersing the sample in a sodium sulfide solution at ambient temperature. The CBD procedure uses a solid electrochemical reaction for the synthesis of Ag_2S. Secondly, EBI is a process that uses high energy electrons to treat objects for various applications, such as the fabrication of TERS tips. The creation of an Ag bridge from the applying of voltage and Ag ions in Ag_2S diffused on the surface. And Ag ions received electrons, then a Ag bridge was formed. CBD combined with EBI methods gives high reproducibility and an orderly arrangement of nanostructures, which are promising methods to fabricate Ag tips.

In this work, CBD and EBI conditions with the fabrication of AgNF tips were optimized to obtain good TERS probes. To avoid natural oxidation, Ag tips were dipped in aqueous Na_2S solution after Ag was instantly coated on the tip. Moreover, we studied the irradiation time in the EBI process.

2. Materials and Methods

2.1 Materials

Sodium sulfide nonahydrate ($Na_2S{\cdot}9H_2O$) (98%) was purchased from Carlo Erba Reagenti. Ethyl alcohol (C_2H_5OH) (95%) was manufactured by Merck. All chemicals and solvents are analytical grade without further purification. Ag and Ti targets (99.99%) with a 3-inch diameter and 0.25-inch thickness were purchased from Kurt J. Lesker Company. N_2 (95.5%) and Ar (99.999%) gases were obtained from S.I. Technology Co., Ltd. and Praxair Co., Ltd., respectively. The AFM probes were purchased from Applied NanoStructures, Inc.

2.2 Instruments

For Ag coating on tips, radio frequency (RF) magnetron sputtering with a thickness monitor (INFICON SQM-160) manufactured by Oerlikon, Germany was used. The morphology of the Ag_2S and AgNF tips was examined using a field emission scanning electron microscope (FE-SEM, Versa 3D DualBeam). The energy dispersive X-ray spectrometer (EDS) was used for elemental composition analysis and elemental mapping, manufactured by Field Emission. The fabricated AgNF tip was also examined using a Raman spectrometer (NT-MDT NTEGRA SPECTRA) equipped with an IX71 Olympus confocal microscope with a 100x objective lens.

2.3 Synthesis

2.3.1 Synthesis of Ag tips

Ag tips were prepared by a sputtering technique. The AFM tips were used as a substrate. The Ag thickness was varied between 400 and 600 nm by adjusting in sputtering process. The FE-SEM was used to characterize the morphology.

2.3.2 Synthesis of Ag_2S tips

The Ag_2S tips were prepared using a CBD technique. Na_2S 0.039 g was dissolved in 100 ml deionized water. After that, the Ag tips from the sputtering technique were dipped into a Na_2S solution for 6 hours to form the Ag_2S tips. The Ag_2S tips were characterized by FE-SEM and EDS mapping to study the surface and chemical compositions of the Ag_2S tips.

2.3.3 Synthesis of Ag nanofilament (AgNF) tip

The AgNF tips were prepared by electron beam irradiation (EBI) in SEM. To study AgNF, each AgNF tip was fabricated under different conditions of accelerated voltage (5 and 20 kV), irradiation time from EBI (10 and 20 minutes), and different Ag thicknesses from the sputtering technique. The AgNF density and length were calculated using ImageJ (ver. 1. 51j8, Wayne Rasband, National Institutes of Health, USA).

3. Results and Discussion

3.1 Effect of accelerated voltage on AgNFs

The AFM tips were prepared with an Ag film thickness of 400 nm by RF magnetron sputtering. The Ag_2S tip was synthesized using the CBD technique. Figs. 1a and b demonstrate SEM images of the Ag_2S tip after irradiation by the EBI process using accelerated voltage at 5 and 20 kV, respectively. Using accelerated voltage of 5 kV, it was found that AgNFs were not formed on the Ag_2S tips. Therefore, the electron on surface of Ag_2S tips does not react with Ag^+, resulting in no forming of AgNFs on Ag_2S tips at 5 kV. At 20 kV, as shown in Fig. 1b, it was found that AgNFs formed at the Ag_2S tips due to the high electron accelerated voltage. Ag^+ in β-Ag_2S can move to the very end of the Ag_2S tip apex to receive electrons to form the Ag nanofilaments.

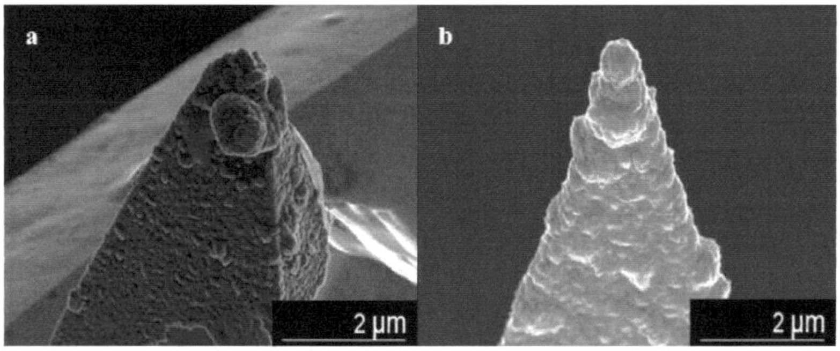

Figure 1. SEM images of the Ag_2S tip after irradiation with different accelerated voltages of (a) 5 and (b) 20 kV.

3.2 Effect of Ag thickness on AgNFs *fabrication*

Ag tips with a Ag thickness of 400 and 600 nm were prepared by RF magnetron sputtering. The Ag tips were immersed in a 0.5 mM sodium sulfide solution for 6 hours. The morphology of the AgNF tip fabricated from the tip with a Ag thickness of 400 and 600 nm was studied by SEM using accelerated voltage at 20 kV, as shown in Figs. 2a and b, respectively. The AgNFs fabricated from the tips with the 600 nm Ag thickness emerged faster than those coated with an Ag thickness of 400 nm. Figs. 2a and b show SEM images of AgNFs terminated at 12 and 18 minutes, respectively. The AgNFs were fabricated from 400 and 600 nm thick Ag tips with 0.184 ± 0.050 and 0.502 ± 0.050 µm diameter. The length of AgNFs on the tip apex of 400 and 600 nm thickness was 0.568 ± 0.040 and 0.702 ± 0.040 µm, respectively. Moreover, the highest density of AgNF was 52.5×10^9 NF/cm^2. The AgNFs with a 400 nm Ag thickness were rod-shaped, while the AgNFs with the 600 nm Ag thickness were cone shaped. They were affected by the jump in band gap energy of Ag^+ in Ag_2S tips at different Ag thicknesses. Therefore, the AgNFs at different Ag thicknesses provided different shapes.

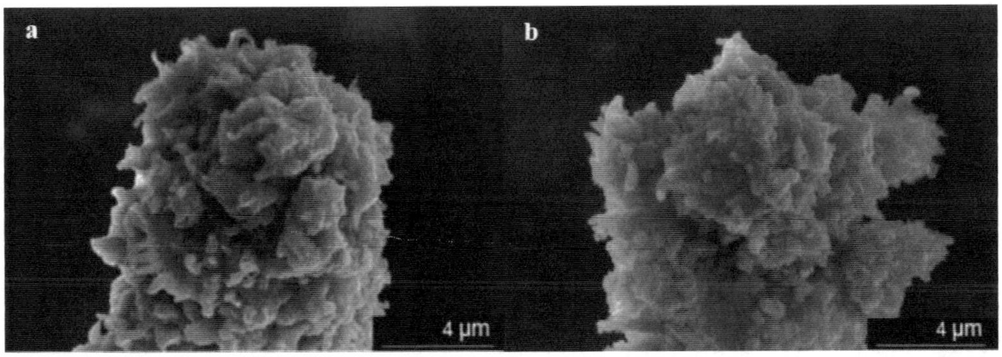

Figure 2. SEM images of AgNF tips fabricated from different Ag thicknesses of (a) 400 and (b) 600 nm.

3.3 Effect of irradiation time on AgNFs fabrication

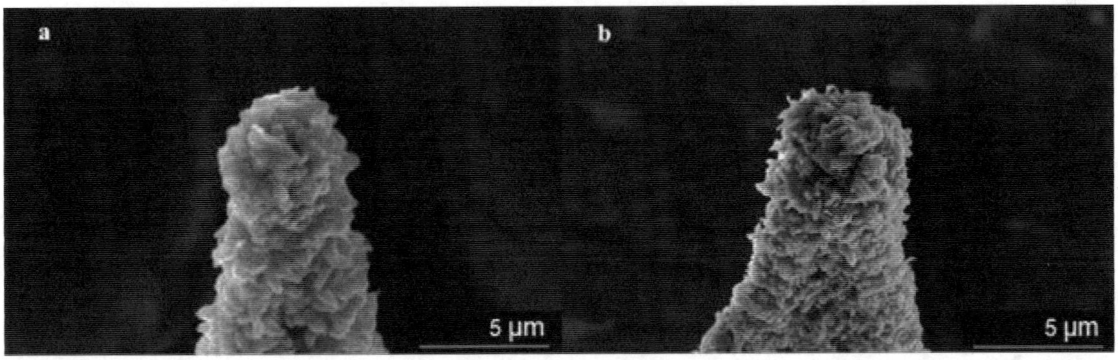

Figure 3. SEM images of the AgNF tip fabricated with an irradiation time of (a) 10 and (b) 20 minutes.

Ag_2S tips of 400 nm Ag thickness are shown in Fig.3. The Ag_2S tip was irradiated with an electron beam for 10 minutes, while, in Fig. 3b, it was irradiated for 20 minutes. The AgNFs were formed rapidly at the tip apex due to the high accelerated voltage conditions of 20 kV and sufficient irradiation time. Moreover, there was a high Ag^+ to produce AgNFs, bringing about the opportunity of Ag^+ in β-Ag_2S which led to a rapid reaction on the surface of the Ag_2S tips. The density of AgNFs after 10 and 20 minutes of the EBI process was 40.8×10^9 and 46.4×10^9 NF/cm^2, respectively. The result shows that the number of AgNFs with an irradiation time of 10 minutes was less than that of 20 minutes. Because less irradiation time resulted to diffuse of Ag^+ ions in β-Ag_2S, they had a shorter time to react with electrons on the surface. In addition, the density related to the gap between AgNFs on tips. Therefore, irradiation time affects the gap of the AgNFs tips.

4. Conclusions

In conclusion, the fabrication of the AgNF tip was successfully carried out using the solid electrolyte of silver sulfide, which can be synthesized *via* chemical bath deposition. To form Ag_2S tips, Ag tips were achieved by immersing them in the 0.5 mM sodium sulfide solution for 6 hours. An accelerated voltage of 20 kV was found to be suitable for AgNF fabrication prepared by electron beam irradiation and 20 minutes of irradiation time. The factors of Ag thickness, accelerated voltage, and irradiation time affect the occurrence of AgNFs, such as different shapes, sizes, and densities. This study shines a light on using the AgNF tip as a tip in tip-enhanced Raman spectroscopy.

Acknowledgments

The authors are grateful to the National Nanotechnology Center (NANOTEC), National Science and Technology Development Agency (NSTDA) for the financial support of this work under grant numbers P1851717 and P2150820. The scholarship is also supported by the Thailand Graduate Institute of Science and Technology (TGIST) under contract number SCA-CO-2561-4373-TH and the Thailand Science Research and Innovation Fund Chulalongkorn University (CU_FRB65_food (14)_122_23_52).

References

[1] Mochalov K E and Solovyova D O et al 2020 *Tech. Phys. Lett.* **46** 1084–1087
[2] Agressott E V H, de Moura T A et al 2020 *Vib. Spectrosc.* 103104
[3] Liu Y and Zhao Y et al 2019 *Spectrochim. Acta A Mol. Biomol. Spectrosc.* **219** 539–546.
[4] Wang R and Kurouski D 2018 *J. Phys. Chem.* C **122** 24334–24340
[5] Liu H and Huang Z et al 2015 *J. Cryst. Growth* **419** 20–24
[6] Micco A and Ricciardi A et al 2015 *Sci. Rep.* **5**
[7] Sivashanmugan K and Liao J D et al 2015 *Sens. Actuators B Chem* **206** 415–422.
[8] Ma C and Chen X et al *CrystEngComm* **20** 2227–2232

Flexible Solution-Processed Agar Material for Resistive Switching Memory

B J Chen[1], Y C Chang[2], J C Jian, H J Liu

[1]Department of Engineering Science, National Cheng Kung University, Tainan 70101, Taiwan
[2]Department of Engineering Science, National Cheng Kung University, Tainan, 70101, Taiwan

*E-mail: christina780712@gmail.com

Abstract. Pure agar (PA)-based resistive random access memory (RRAM) devices using simple solution process method were demonstrated. The memory properties of the investigated PA-based resistive layers are determined by the relationship between the physical properties of the PA thin films and the RRAM performance. The memory properties of PA films with 2 different bottom substrates (glass and PET) and ITO as bottom electrodes were also investigated to explore the differences between rigid and flexible substrates. The RRAM fabricated from PA film and PET substrates exhibit good uniformity acceptable retention over 10^4 s, and high ON/OFF ratio (>10^4) at room temperature. According to the fitting results, the RRAM made of PA thin film conforms to the Space-charge-limited current model. This work helps to understand the resistive switching and maintain stable operations of Agar-based RRAM, which means a lot for increasing the application of agar in flexible devices.

1. Introduction

Nowadays, electronic products are often manufactured from potentially toxic materials, which also pollute water and soil. Due to the increasing of environmental awareness, biomaterials become strong candidate for alternatives to fossil fuel materials because of their properties, such as inexhaustible, eco-friendly processability, and mechanical flexibility. As a result, electronics consisting of biodegradable biomaterials may help reduce the accumulation of permanent solid waste and make our environment better. So far, renewable materials from animals, plants, and bacteria that have been used in electronics include pectin [1, 2], proteins [3-5], silk [6, 7] Konjac Glucomannan Material [8], and gelatin [9-11]. Among various natural materials, agar is one of important biopolymer due to its good film-forming ability, flexibility, transparency, biocompatibility, and biodegradability. These properties can expand the application range of Agar, such as food and biomedicine [12-14]. Agar is a suitable biomaterial for a low-cost solution process because of its plasticity and excellent film formation. The prepared Agar film can be stored under atmospheric conditions and has a transparent character in visible light [13]. Thus, agar is appropriate for transparent applications. However, the application of Agar-based electronic devices has not been widely discussed.

In this study, agar was used as an insulating layer based on ITO/glass and ITO/ PET substrates for RRAM without any further purification [1, 2, 8]. Recent studies show that flexible substrates and biomaterials are less used together in RRAM [15-17]. Therefore, the device performance of Agar-based

RRAM fabricating on different substrates was investigated, especially the differences between rigid and flexible substrates. The flexible RRAM device with Al/Agar/ITO/PET structure has invisible bipolar resistive switching characteristics, a good ON/OFF ratio over 10^4, excellent uniformity of less than 40%, and stable retention over 10^4 s. These memory properties suggest that flexible devices fabricated from Agar solutions and PET substrates have great potential for bio-based RRAM devices.

2. Experiment

2.1. Sample preparation

Agar powder was mixed with DI water to prepare pure agar solution (PA). The PA concentrations were adjusted into 0.2, 0.6, and 1.0 % wt, which were denoted as PA02, PA06, and PA10, respectively. After surface cleaning, the agar solution was spun onto the cleaned ITO/glass and ITO/PET substrates. Subsequently, the devices were baked at 60°C for 30 minutes. Finally, the Al electrodes with thicknesses of around 200 nm were deposited on top of the agar thin film by radio frequency magnetron sputtering. The RRAM structure is shown in Fig 1, the top electrode is Al, the isolator is PA06 thin film, the bottom electrode is ITO.

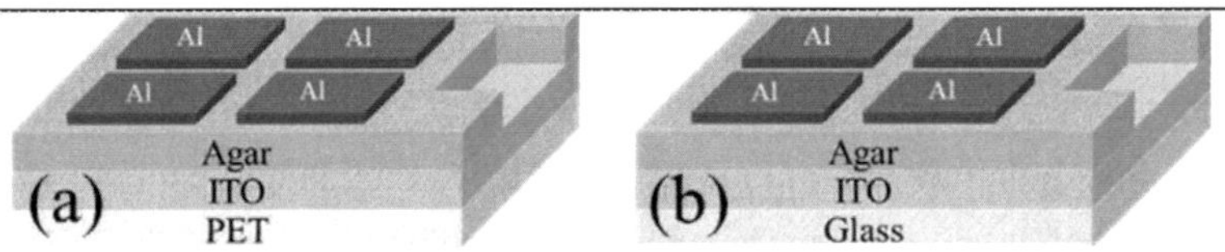

Figure .1 (a) Agar-based RRAM structure on glass substrates and (b) PET substrates

2.2. Physical analysis

We characterized the surface morphology of the obtained PA/ITO samples by Atomic force microscopy (AFM) analysis. Fourier-transform infrared (FTIR) spectroscopy was carried out using a Vertex 80v and Tensor 27 (Bruker) in the range of 4000– 800 cm^{-1}. The percentage of C、N and O elements were confirmed by X-ray photoelectron spectroscopy (XPS). The transmittance of the device was verified by ultraviolet and visible absorption spectroscopy (UV–vis) with a Perkin-Elmer Lambda 35 spectrophotometer. The electrical properties were measured with a Keithley 2636B system source meter.

3. Result and discussion:

3.1. Physical analysis

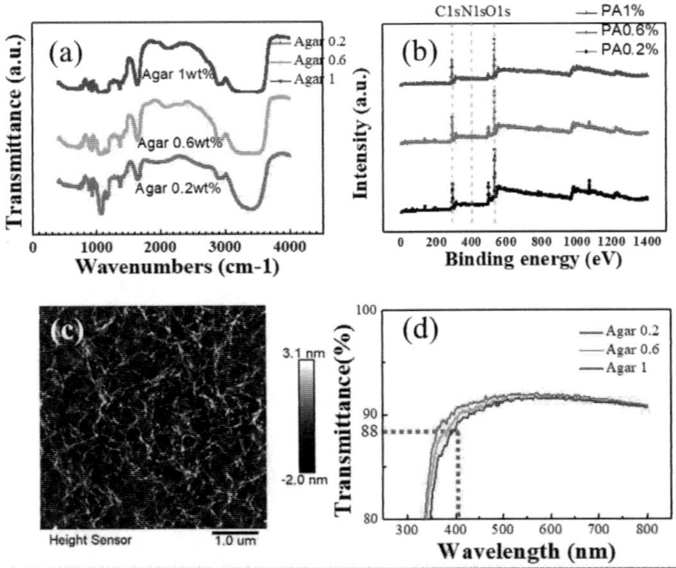

Figure .2 (a) FTIR analysis of different concentrations of agar solution. (b) XPS analysis of different concentrations of agar. (c) The AFM analysis result of PA06. (d) UV-Vis analysis of different agar concentrations

To identify the functional groups of agar, we used FTIR analysis to observe the state of energy absorption Fig 2. (a). The peaks of energy absorption occurred at 3400, 2900, 1643, 1375, 1070, and 930 cm^{-1}. The band region of 3400 cm-1 corresponds to OH bond stretching vibration. The band at 2900 cm^{-1} is $-OCH_3$ for energy absorption, 1643 cm^{-1} is $-NH_2$ and –C=O bending vibration, 1375 cm^{-1} is sulfate for energy absorption, 1070 $cm^{-}1$, and 930 cm^{-1} is related to the stretching vibration of CO in 3,6-anhydro-galactose [12-14] . We could observe the significant change in the energy absorption of functional groups across the three concentrations of agar.

Fig 2. (b) shows the XPS results of different concentrations of Agar films. The C, O, and N composition ratios of PA02 and PA10 were 57%, 41.4%, 1.6%, and 59.1%, 39.6%, 1.3%, respectively. The element percentage of C, O, and N in the PA10 were 61.2%, 36.4%, and 2.4%. The root means square roughness of Agar surface also estimated by AFM analysis. The surface roughness of PA06 is 0.535 nm. Fig 2. (c) shows the root means square roughness of Agar surface also estimated by AFM analysis. The surface roughness of PA06 is 0.535 nm.

As shown in Fig 2. (d), the transmittance of the RRAM made of three different concentrations of PA films reached 88% when the wavelength was greater than 400 nm, indicating that the solution could be made into a transparent device. In addition, it can be observed that as the concentration of agarin is greater, the transparency is smaller under the same wavelength.

IOP Conf. Series: Materials Science and Engineering 1250 (2022) 012006 doi:10.1088/1757-899X/1250/1/012006

3.2. Electrical properties

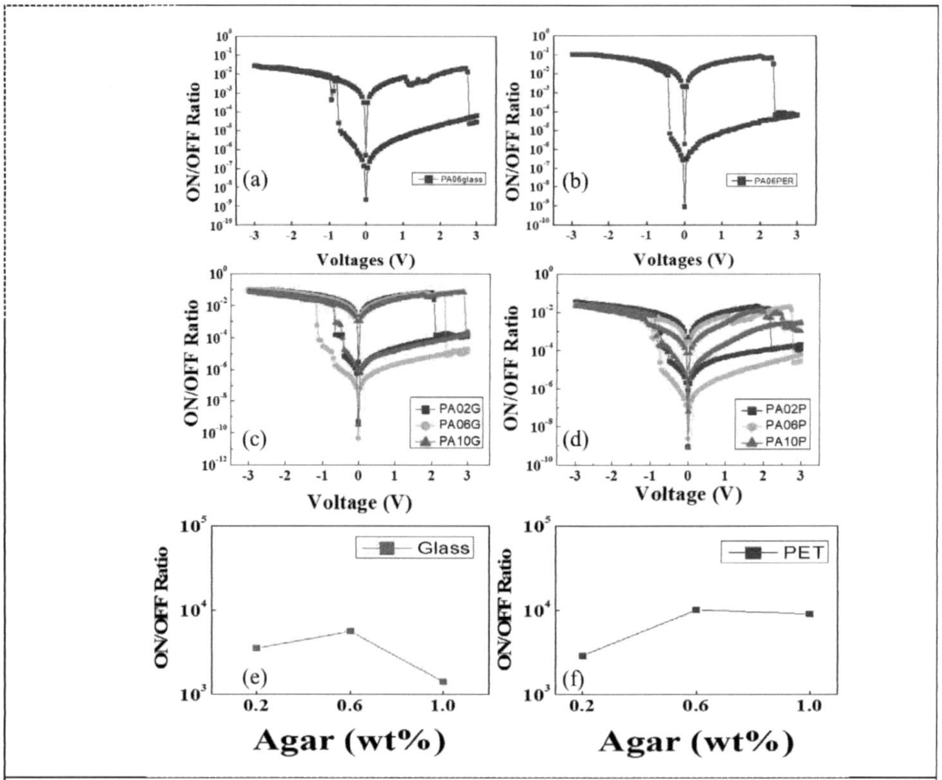

Figure. 3 (a) *I-V* curve of PA06/glass. (b) *I-V* curve of PA06/PET. (c) *I-V* curve of 0.2,0.6,1.0% PA on the glass substrates. (d) *I-V* curve of 0.2,0.6,1.0% PA on the PET substrates. (e) ON/OFF ratio of 0.2,0.6,1.0% PA on the glass and (f) PET substrates.

Fig 3. (a) and Fig 3. (b) show the electrical properties of PA06 under PET and Glass, both PA06/PET and PA06/glass exhibit bipolar resistive switching behavior. It also shows the on/off ratio and set voltage (V_{set})/reset voltage (V_{reset}) of the PA06/PET and PA06/glass. The on/off ratio of PA06/glass is 5.66×10^3, and the on/off ratio of PA06/PET is 1.06×10^4 which is one order larger than PA06/glass. From Fig 3. (c) and Fig 3. (d) the V_{set} / V_{reset} of PA60/PET and PA06/glass are -0.45V/2.4V and -0.8V/2.8V, respectively. The electrical properties of PA02, PA06, and PA10 on different substrates were compared. Fig 3. (e) shows the ON/OFF ratio of PA02, PA06 and PA10 on different substrates, we could learn that the on/off ratio of PA06 is larger than that of PA02 and PA10 whether on ITO/PET or ITO/Glass substrate devices. Fig 4. (a) is the cumulative probability of the high resistance state (HRS) and low resistance state (LRS) current values for the PA06/glass memory device under a read bias of 0.1 V.

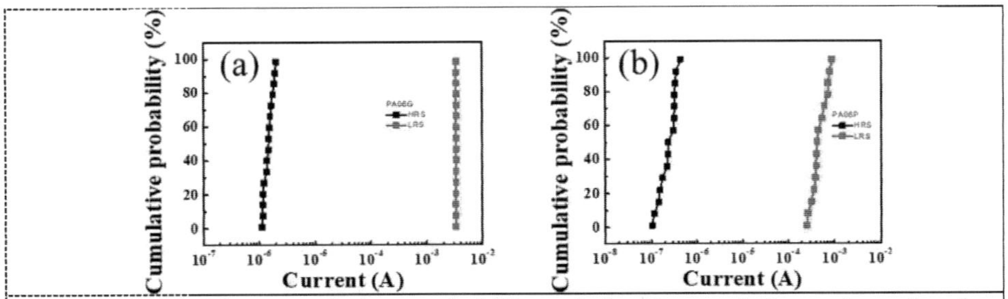

Figure .4 (a) The cumulative probability of the I_{HRS} and I_{LRS} values for the PA06/glass memory device under a read bias of 0.1 V. (b) The cumulative probability of the I_{HRS} and I_{LRS} values for the PA06/PET memory device under a read bias of 0.1 V.

The I_{HRS} of PA06/glass is 1.49×10^{-6} and the CV value is 19.09%, the I_{LRS} is 3.32×10^{-3} and the CV value is 0.67%. Fig. 4 (b) is the cumulative probability of the I_{HRS} and I_{LRS} values for the PA06/PET memory device under a read bias of 0.1 V. The I_{HRS} of PA06/glass is 2.88×10^{-7} and the CV value is 44.14%, the I_{LRS} is 5.02×10^{-4} and the CV value is 37.97% . As a rigid substrate, glass substrates have better stability. Although the stability of PET substrates is not well as glass substrates, it is still good enough. Besides, the PET substrates have a larger ON/OFF ratio and smaller V_{set} / V_{reset}. With the above advantages mentioned, agars based on PET substrates can be applicated in a wider range of the flexible devices.

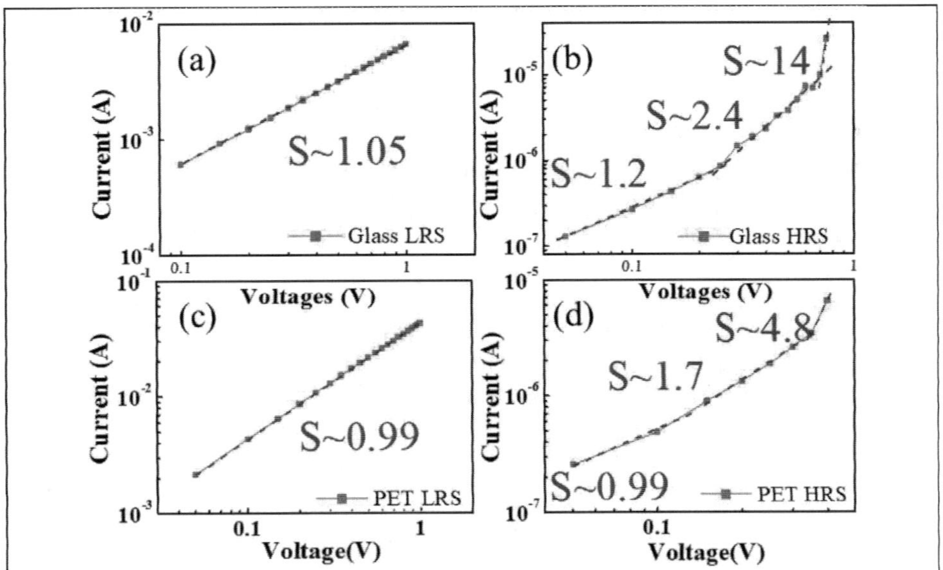

Figure .5 (a) Double logarithmic plots of the low resistance state (LRS) and (b)high resistance state (HRS) of the PA06/glass memory device. (c) Double logarithmic plots of the LRS and (d) HRS of the PA06/PET memory device.

To further clarify on the conduction mechanisms of agar RRAM, the obtained typical *I–V* characteristics were investigated, as shown in Fig. 5. The *I–V* relationship in the LRS (ON state) exhibits ohmic conduction behavior for PA06/glass and PA06/PET with a slope of about 1.05, and 0.98, respectively. The Space-charge-limited current (SCLC) model is used to describe three different conductive regions:

(1) the low-voltage region where the IV curve exhibits linear behavior , the slope of which is around 1.2 for PA06/glass and about 0.98 for PA06/PET ($I \propto V$), corresponding to the ohmic conduction mechanism. (2) the transition region where the current increases non-linearly with the squared voltage dependence ($I \propto V^2$) and the slope increases, the PA06/glass slope is around 2.4, the PA06/PET slope is around 1.66, which is close to *Child's law* with a slope of 2, and (3) a region marked by a sharp increase in current, where the PA06/glass slope is approximately 14 and the PA06/PET slope is approximately 4.85. The findings on the *I-V* properties suggest that the different conduction behaviors of Agar-based RRAM devices are controlled by the formation of SCLC and localized filament paths.

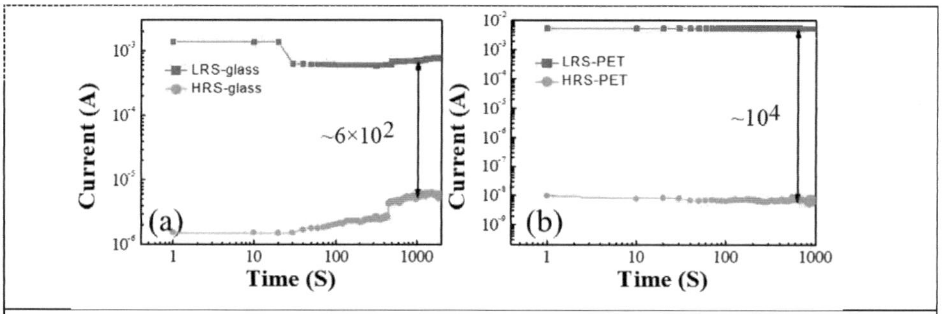

Figure .6 Retention property of (a) the PA06G and (b) the PA06P memory device at room temperature.

Fig. 6 shows the memory retention of PA06/glass and PA06/PET. The ON/OFF ratio of the PA06/PET memory device exceeds 10^4, and no obvious attenuation is observed within 10^4 s. While the ON/OFF ratio of the PA06/glass memory device is around 6×10^2 at 10^4 s, indicating that PA06/PET has better memory retention.

References

[1] Chang, Y. C., Jian, J. C., Chuang, M. Y., Hsu, Y. L., Huang, W. Y., & Young, S. J.. Metal and Carbon Filaments in Biomemory Devices through Controlling the Al/Apple Pectin Interface. *ACS Applied Electronic Materials* 2020, 2(9), 2798-2805.

[2] Lin, K. W., Wang, T. Y., & Chang, Y. C.. Impact of Top electrodes on the nonvolatile resistive switching properties of citrus thin films. *Polymers*, 2021 13(5), 710.

[3] Moudgil, A.; Kalyani, N.; Sinsinbar, G.; Das, S.; Mishra, P. S-Layer Protein for Resistive Switching and Flexible Nonvolatile Memory Device. *ACS Appl. Mater. Interfaces* 2018, 10, 4866– 4873.

[4] Raeis-Hosseini, N.; Park, Y.; Lee, J. S. Flexible Artificial Synaptic Devices Based on Collagen from Fish Protein with Spike-Timing-Dependent Plasticity. *Adv. Funct. Mater.* 2018, 28, 1800553.

[5] Zhu, J. X.; Zhou, W. L.; Wang, Z. Q.; Xu, H. Y.; Lin, Y.; Liu, W. Z.; Ma, J. G.; Liu, Y. C.; Liu, Y. C. Flexible, Transferable and Conformal Egg Albumen Based Resistive Switching Memory Devices. *RSC Adv.* 2017, 7, 32114– 32119.

[6] Qu, B.; Lin, Q.; Wan, T.; Du, H.; Chen, N.; Lin, X.; Chu, D. Transparent and Flexible Write-Once-Read-Many (WORM) Memory Device Based on Egg Albumen. J. Phys. D: *Appl. Phys.* 2017, 50, 315105.

[7] Zhou, Z.; Shi, Z.; Cai, X.; Zhang, S.; Corder, S. G.; Li, X.; Kaplan, D. L. The Use of Functionalized Silk Fibroin Films as a Platform for Optical Diffraction-Based Sensing Applications. *Adv. Mater.* 2017, 29, 1605471.

[8] Chang, Y. C., Liu, H. J., Jian, J. C., & Hsu, Y. L.. Solution-Processed Natural Konjac Glucomannan Material for Resistive Switching Memory. *ACS Applied Electronic Materials.* 2021.

[9] Chang, Y. C., & Wang, Y. H.. Resistive switching behavior in gelatin thin films for nonvolatile

memory application. *ACS applied materials & interfaces*, 2014, *6*(8), 5413-5421.
[10] Chang, Y. C., & Wang, Y. H.. Solution-processed Al-chelated gelatin for highly transparent non-volatile memory applications. *Applied Physics Letters*, 2015, 106(12), 34_1.
[11] Chang, Y. C., Jian, J. C., Hsu, Y. L., Huang, W. Y., & Young, S. J.. A Green Strategy for Developing a Self-Healing Gelatin Resistive Memory Device. *ACS Applied Polymer Materials*, 2020, 2(11), 5318-5326.
[12] Araki, C.. Structure of the agarose constituent of agar-agar. *Bulletin of the Chemical Society of Japan*, 1956, *29*(4), 543-544.
[13] Bertasa, M., Dodero, A., Alloisio, M., Vicini, S., Riedo, C., Sansonetti, A.& Castellano, M. et al. Agar gel strength: A correlation study between chemical composition and rheological properties. *European Polymer Journal*, 2020, 123, 109442.
[14] Armisen, R., & Galatas, F. Production, properties and uses of agar. Production and utilization of products from commercial seaweeds. *FAO Fish. Tech. Pap*, 1987.
[15] Lin, C. C., Chen, Y. D., & Lin, N. C.. Graphene oxide based device for flexible RRAM application. In *2013 International Symposium on Next-Generation Electronics* . *IEEE,* 2013, February, . 396-398.
[16] Chou, K. I., Cheng, C. H., Zheng, Z. W., Liu, M., & Chin, A.. Ni/GeOx/TiOy/TaNNi/GeOx/TiOy/TaN RRAM on Flexible Substrate With Excellent Resistance Distribution. *IEEE electron device letters*, 2013, 34(4), 505-507.
[17] Lee, M. J., Lee, C. B., Kim, S., Yin, H., Park, J., Ahn, S. E., ... & Park, Y.. Stack friendly all-oxide 3D RRAM using GaInZnO peripheral TFT realized over glass substrates. *In 2008 IEEE International Electron Devices Meeting IEEE*. 2008, December, pp. 1-4.

Conversion of Steel Converter Slag into Anion Removing Agent via Alkali Fusion

Takaaki Wajima

Department of Urban Environment Systems, Graduate School of Engineering, Chiba University, 1-33, Yayoi-cho, Inage-ku, Chiba 263-8522, Japan

wajima@tu.chiba-u.ac.jp

Abstract. Steel converter slag, a by-product of iron- and steel-making plants, was converted into a product containing hydrocalumite-like compound using the alkali fusion method. Most of the crystalline phase in the converter slag was converted into a soluble phase by the fusion method, and could be transformed into a product which included hydrocalumite-type crystals. The ability of the product to remove BO_3^-, F^-, PO_4^{3-}, AsO_2^-, AsO_4^{3-}, SeO_3^{2-} and SeO_4^{2-} from aqueous solution is higher than that of raw slag, especially for BO_3^-, AsO_2^-, AsO_4^{3-} and SeO_3^{2-}.

1. Introduction

Steel converter slag is produced in large quantities worldwide as a by-product in primary steelmaking process; for example, approximately 90 kg of slag is produced per ton of steel [1]. In the EU alone, with a primary steel production of approximately 110 million tons, 10 million tons of slag is produced annually [2]. In Japan, approximately 98 million tons of crude steel is produced annually, thus generating approximately 13 million tons of steel converter slag as a by-product [3]. Steel slag is utilized in construction industry to replace cement; however, cement can be replaced only by 10–20% with steel slag because of the latter's low hydration reactivity [4]. Converter slag is also used as an aggregate; however, volume stability problems owing to the presence of free lime are common [5]. Thus, the use of steel slag in these applications is limited, and a major portion of converter slag is disposed as a backfill material or in landfills. Therefore, the development of value-added products from slag has become an important issue to ensure sustainability of the converter process.

Hydrocalumite is an anionic clay mineral belonging to the layered double hydroxide (LDH) family, and is expressed as $3CaO \cdot Al_2O_3 \cdot CaX_{2/m} \cdot nH_2O$ (X: univalent or bivalent anion, n: 20 or less). LDHs are lamellar mixed hydroxides containing positively charged main layers and they undergo anion exchange [6]. In recent years, many studies have been devoted to investigating the ability of LDHs to remove harmful oxyanions such as arsenate, chromate, phosphate, etc. from contaminated water [6]. In our previous studies, we converted blast furnace slag into a hydroxide-based adsorbent that contains calcite and Ca-Al-type hydrocalumite through alkali fusion, and confirmed that this material is an adsorbent with high affinity for the removal of oxi-anions from aqueous solutions [7-9]. However, little information can be available on the conversion of steel converter slag into the adsorbent including the Ca-Fe-type hydrocalumite with the bility of anion removal ability.

The purpose of this study is to ascertain whether conversion of the steel converter slag into a hydrocalumite-type compound (mainly CaFe-LDHs) is possible using the alkali fusion method. The crystalline phase of the slag was first converted into a soluble phase by alkali fusion, and then used to synthesize the hydrocalumite-type compound. In addition, the ability of the product to remove anions from solution was examined for possible application in wastewater treatment.

2. Experimental

2.1. Steel converter slag

The steel converter slag used in this study was received from a steel-making plant in Japan. The converter slag was ground in a mill and sieved to a particle size of less than 1 mm. The chemical and mineralogical compositions of the raw slag were determined by X-ray fluorescence spectrometry (XRF; Epsilon1, Malvern Pananalytical, Japan) and X-ray diffraction (XRD; MiniFlex600, Rigaku, Japan), respectively (Table 1 and Fig. 1). The raw slag was mainly composed of CaO (48.1%), Fe_2O_3 (20.0%), and SiO_2 (13.1%), found as calcite ($CaCO_3$), magnetite (Fe_3O_4), and larnite (Ca_2SiO_4), respectively. Other oxides such as Al_2O_3, MgO, MnO, P_2O_5, TiO_2, SrO, Cr_2O_3, V_2O_5, and Na_2O were also present in smaller amounts.

2.2. Synthesis

Chemical conversion was carried out as follows; 20 g of the slag was mixed with 32 g of NaOH powder, and ground to obtain a homogeneous mixture. This mixture was heated in a nickel crucible in air at 600°C for 6 h. The resultant material was cooled to room temperature in air and ground again to obtain the fused material. The fused material was added to 200 mL of distilled water, and stirred with a stirrer (600 rpm) at room temperature for 24 h. The solid was filtered, washed with distilled water and dried in a drying oven at 60°C overnight to obtain the final product. The chemical composition of the raw slag and the product were determined by XRF. Characterization of the slag, fused material and product was performed by XRD and scanning electron microscopy (SEM) (Hitachi, S-4500), respectively. The total gravity/differential thermogravimetric analysis (TG - DTA) by heating at a heating rate of 10 °C/min in an air atmosphere for raw slag and the product was performed using a differential thermogravimetric analyzer (STA6000, Perkin Elmer).

2.3. Anion removal

The ability of the raw slag and the product to remove anions from solution was evaluated. 1 mM solutions with BO_3^- (B), F^- (F), PO_4^{3-} (P), AsO_2^- (As(III)), AsO_4^{3-} (As(V)), SeO_3^{2-} (Se(IV)) and SeO_4^{2-} (Se(VI)) were prepared from $Na_2B_4O_7{\cdot}10H_2O$, NaF, $Na_2HPO_4{\cdot}12H_2O$, $NaAsO_2$, $Na_2HAsO_4{\cdot}7H_2O$, Na_2SeO_3 and Na_2SeO_4, respectively. 0.1g of the slag or the product was added to 20 mL of the solution in a 50 mL centrifuge tube, and shaken for 1 h using a reciprocal shaker. After shaking, the solution was centrifuged at 3000 rpm for 10 min. The concentrations of F^- and PO_4^{3-} in the supernatant were determined by ion meter (TOKO, TiN-5101) and molybdenum blue method, respectively, and those of AsO_2^-, AsO_4^{3-}, SeO_3^{2-} and SeO_4^{2-} were determined using inductively-coupled plasma atomic emission spectroscopy (PerkinElmer, Avio500). The percentage anion removal (R) was calculated from the difference between the initial (C_0) and final solution concentrations (C) using following equation:

$$R = \frac{(C_0 - C) \cdot 100}{C_0} \quad (1)$$

3. Results and discussion

Figure 1 shows the XRD patterns obtained for the (a) slag, (b) fused material and (c) product. The slag was composed of the crystalline phases, larnite, magnetite and calcite (Fig. 1 (a)). After fusion, most of the crystalline phases were converted to soluble sodium salts, such as sodium iron oxide, natrite, natrosilite and sodium silicate (Fig. 1 (b)), and finally, the product which included the Ca-Fe-type hydrocalumite and portlandite, was synthesized.

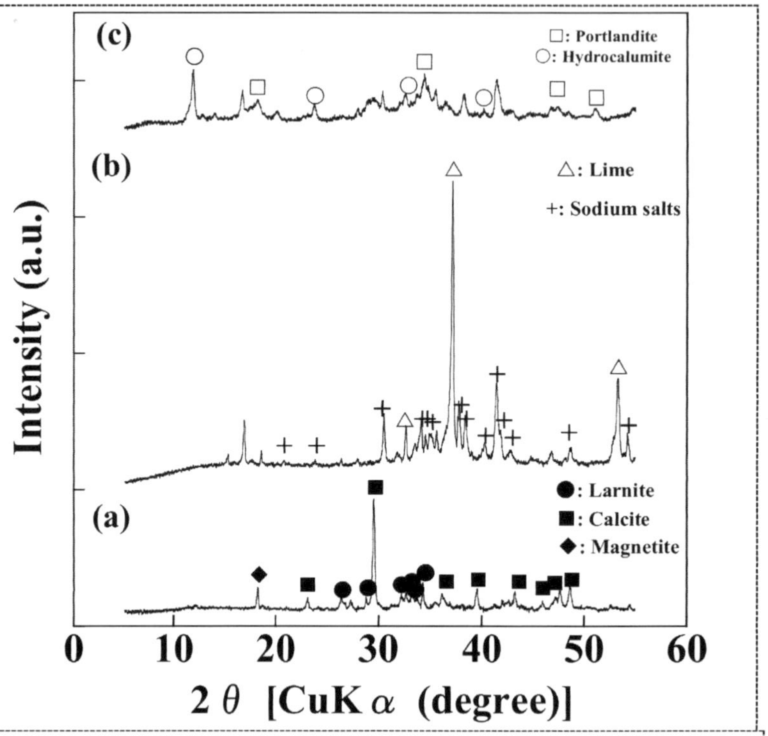

Figure 1. XRD patterns of (a) raw slag, (b) fused material and (c) product.

Figure 2 shows the SEM micrographs obtained for the (a) raw slag, (b) fused material and (c) product. Although the slag is composed of fine particles (Fig. 2 (a)), the fused material consists of large particles with a melted surface resulting from the formation of sodium salts by alkali fusion (Fig. 2 (b)). After synthesis, thin regular hexagonal crystals of the hydrocalumite-type compound, were confirmed to be present in the product (Fig. 2 (c)).

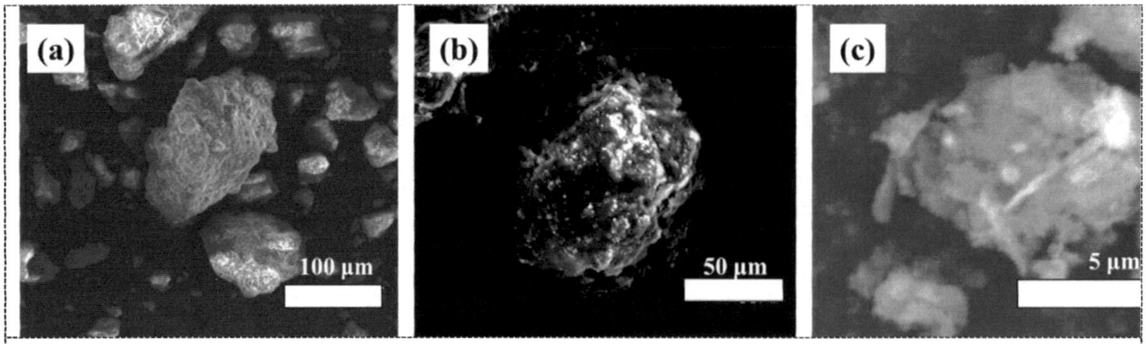

Figure 2. SEM photos of (a) raw slag, (b) fused materials and (c) product.

Table 1 shows the chemical composition of the slag and product. The slag contains higher amounts of CaO (61.3 wt %) and lower amounts of Fe_2O_3 (23.2 wt %) and Al_2O_3 (1.6 wt %) than the product (56.4, 26.7 and 2.1 wt %, respectively). This is most likely caused by the change of main phases from the synthesis of portlandite and the hydrocalumite-type compound.

Table 1. Chemical compositions of raw slag and product.

Oxide (%)	Raw slag	Product
CaO	61.3	56.4
MgO	1.3	1.9
Fe_2O_3	23.2	26.7
Al_2O_3	1.6	2.1
SiO_2	6.3	5.7
TiO_2	0.5	0.7
MnO	3.8	4.4
SrO	0.1	0.1
P_2O_5	0.7	0.8
Cr_2O_3	0.2	0.4
V_2O_5	0.1	0.1
Na_2O	0.1	0.1

The slag is converted into a soluble material by alkali fusion, and the product including hydrocalumite-type compound can then be synthesized from the fused material at room temperature.

Figure shows the anion removal efficiencies for the slag and product. The supernatant pH after the removal tests was approximately 12 regardless of the pH of the original solution. It is noted that both raw slag and product indicate 100 % removal for PO_4^{3-}. Raw slag has the low ability to remove BO_3^- (2.3%), F^- (23.8%), AsO_2^- (13.3%), AsO_4^{3-} (31.7%), SeO_3^{2-} (4.8%) and SeO_4^{2-} (1.0%), owing to its compounds containing Ca and Fe, while the product has the higher ability to remove BO_3^- (26.4%), F^- (41.1%), AsO_2^- (57.7%), AsO_4^{3-} (92.3%), SeO_3^{2-} (35.0%) and SeO_4^{2-} (7.1%) than raw slag, especially for BO_3^-, AsO_2^-, AsO_4^{3-} and SeO_3^{2-}. This is believed to occur as a result of the synthesis of the hydrocalumite-type compound.

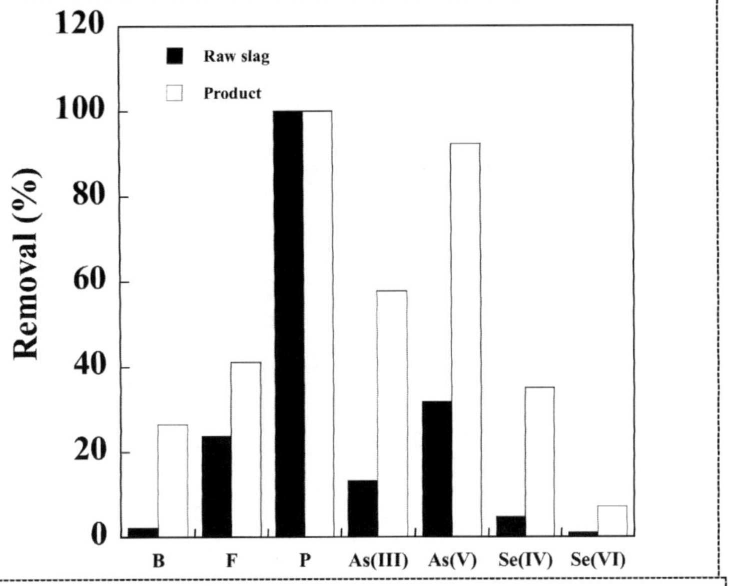

Figure 3. Anion removal using raw slag and product..

4. Conclusion

Steel converter slag is a by-product of iron- and steel-making plants, and the expansion of its utilization is desired. The converter slag including larnite, calcite and magnetite was converted into the fused material with lime and soluble sodium salts, and then the product containing hydrocalumite-like compound and portlandite can be synthesized. The product has the higher removal ability to remove BO_3^-, F^-, PO_4^{3-}, AsO_2^-, AsO_4^{3-}, SeO_3^{2-} and SeO_4^{2-} from aqueous solution than that of raw slag, especially for BO_3^-, AsO_2^-, AsO_4^{3-} and SeO_3^{2-}.

5. References

[1] Fisher L V and Barron A R 2019 *Resour. Consev Recycl.* **146** 244–255.
[2] EUROFER: European Steel in Figures 2018.
[3] The Annual Report of Statistics of Iron and Steel Slag 2019, ed by Nippon Slag Association www. slg.jp/statistics/report.html (accessed 2021-6-1)
[4] Jiang Y, Ling T C, Shi C and Pan S Y 2018 *Resour. Conserv. Recycl.* **136** 187–197.
[5] Brand A S and Roesler J R 2015 *Cem. Concr. Compos.* **60** 1–9.
[6] Goh K H, Lim T T and Dong Z 2008 *Water Res.* **42** 1343-1368.
[7] Wajima T, Oya K, Shibayama A, Sugawara K and Munakata K 2011 *ISIJ Int.* **51** 1179-1184.
[8] Wajima T 2017 *ESTEEMED Acad. J.* **13** 1-17.
[9] Wajima T 2020 *Resour. Process.* **67** 86-93.

AUTHOR INDEX